THE VULNERABILITY OF FIBER OPTIC
A CARVER + SHOCK THREAT ASSESSM
INFORMATION AND COMMUNICATION
TECHNOLOGY SYSTEM INFRASTRUCTURE OF
DOWNTOWN SAN DIEGO, CALIFORNIA

_____

A Thesis

Presented to the

Faculty of

San Diego State University

_____

In Partial Fulfillment

of the Requirements for the Degree

Master of Science

in

Interdisciplinary Studies: Homeland Security

_____

by

Lance William Larson

Spring 2006

DEDICATION

To my mother, who has been a source of encouragement and inspiration to me throughout my life.

To my family, friends, and faculty of San Diego State University's Center for Homeland Security, without whom I could not have completed this thesis.

Experts say terrorism is like lightning. It takes the path
of least resistance to its end.
And, right now, it's easier to blow something up than to
figure out how to damage it by hacking into and
manipulating a computer system.

—Marcus Kempe
Director of Operations
Massachusetts Water Resource Authority

ABSTRACT OF THE THESIS

The Vulnerability of Fiber Optic Networks:
A CARVER + Shock Threat Assessment for the
Information and Communication
Technology System Infrastructure of
Downtown San Diego, California
by
Lance William Larson
Master of Science in Interdisciplinary Studies: Homeland Security
San Diego State University, 2006

It is imperative that fiber optic cabling, arguably the most important of our nation's critical infrastructures, is protected from terrorist attacks and other devastating threats. By utilizing a combined protection effort between national, state, municipal, and private industry, and setting appropriate and enforceable policy for fiber optic lines security and installation, a better strategy for infrastructure protection can be achieved.

Fiber optic cabling, the backbone of information and communication technology (ICT) infrastructure, carries critical data transmissions, phone conversations, and financial transactions from business and government in downtown San Diego to the outside world. Without fully operational fiber optic cabling, business and government operations would grind to a halt.

Agglomeration (the tendency for fiber optic providers to build fiber optic lines closer and closer to one another), and private sector efficiencies, such as building a single trench to encase fiber optic cabling in order to save money, are main contributors to fiber optic vulnerabilities. The City of San Diego also contributes to fiber optic vulnerabilities by relying on fiber optic installation providers to self-regulate, and by promoting the use of shared trenches to minimize construction and pavement disturbances. Finally, location information for fiber optic lines in downtown San Diego city streets is widely available via the Internet, and directly from the City of San Diego using publicly available sources.

Using the CARVER + Shock threat assessment model, originally developed by the military as a targeting assessment tool, ICT infrastructure targets in Downtown San Diego were analyzed and prioritized in a 1500-page threat assessment. Targets with a higher threat score should immediately be evaluated by terrorism planners.

The results of the threat assessment showed several policy shortfalls and lack of policy for fiber optic line installations, several locations where fiber optic cabling agglomerates in critical government and business centers, and several proposed single points of failure.

The results of the study will be disseminated to homeland security agencies and to local governments.

TABLE OF CONTENTS

LIST OF TABLES

LIST OF FIGURES

ACKNOWLEDGMENTS

There are a number of exceptional people without whom this thesis might not have been written and to whom I am greatly indebted and appreciative.

My thesis chair, Dr. Eric Frost, for his constant motivation and praise—at almost all hours of the day and night; Dr. Jeffrey McIlwain for his insight into and endless knowledge of the criminal mastermind; Dr. Theo Addo for his selfless contribution of time editing, beautifying, and converting my thesis into a true scholarly paper.

Steve Price and John Graham for our weekly battles with Open Office spreadsheet formulas. My editor Maggie Jaffe, for her fast editing turnarounds and insightful recommendations. Lili Shim, for her expert knowledge of thesis formatting and proper citations. Norman and Kelli Mjellem, Dr. Sean Gorman, Ranae Larson, Dustin Getchell, and Becca Koenig for reading my first attempts at a coherent thesis. Josh Butler for his civil engineering expertise and knowledge of San Diego city street compositions. Lastly, my friends and family who motivated me to write, even after listening to me talk about my thesis research for hours each week.

CHAPTER 1

INTRODUCTION

On September 11, 2001, the use of commercial passenger airliners as "weapons of mass destruction" immediately caused both government and private industry to re-examine policies for security and disaster recovery. The realization that unconventional weapons can be applied to unconventional targets came as a shock to the entire nation. For the first time in United States history, airborne, seagoing, or underground targets were no longer considered secure and protected from terrorist attacks.

Clearly, homeland security planners needed to reconsider the entire nation's infrastructure, including the miles of fiber optic lines within the United States, as potential terrorist targets. Underground fiber optic cabling carries daily telephone conversations, Internet data, and even sensitive government transactions throughout the country. Information and communication technology (ICT) infrastructure, arguably the most important of our critical infrastructures, is the hardware, software, and web-like network of fiber optic lines, that route communication data from one source to another specific destination. This infrastructure was once considered impervious to terrorist attacks due to its highly diverse pathways and the multitude of disparate private and public networks. Fiber optic lines entrenched across much of the nation during the "Dot Com Boom" were considered to be diverse and "an effective information infrastructure" (Office of the Vice President, par. 8). Is this assumption true?

Thomas Homer-Dixon, an associate professor of political science and director of the Centre for the Study of Peace and Conflict at the University of

1

Toronto, for one, does not believe this assumption: he analogized the nation's fiber optic infrastructure to cars traveling along a highway. When drivers tailgate too closely on the highway, a highly coupled system is created. Mistakes by one driver or a sudden shock from outside the system (such as an animal crossing the road) can cause a chain reaction of accidents, with cars piling on top of one another, essentially choking the system (Dixon, par. 14).

RESEARCH FOCUS: DOWNTOWN SAN DIEGO

San Diego, California, is the seventh largest city in the United States. With seventy miles of coastline and thirty-three different beaches, San Diego is home to 1.2 million culturally diverse residents with a median family income of $59,900. San Diego has over eighty-five cultural institutions, including the world-renown San Diego Zoo in 1,400 acres of Balboa Park alone. San Diego attracts over 5,000 businesses and 75,000 center city employees—who are among the highest percentage of college graduates and PhDs in the United States. With over 9 million square feet of business office space, thriving defense, military, biotechnology, and tourism industries as well as a mild, "Mediterranean-like" climate, it is no wonder that San Diego is an ideal city in which to live and to do business (Center City Development Corporation [CCDC] 1–12).

Bordered to the north by Laurel Street, to the east by Interstate-5, to the south by Commercial Street, and to the west by the Pacific Ocean, the 2.4 square mile area of downtown San Diego is the center for business, government, and tourism. With the lowest business tax of the twenty largest U.S. cities, and with eighteen incorporated areas, it is no surprise that an estimated 175,000 people will work downtown by 2030. To date, over $3 billion in redevelopment funds have been spent for downtown area projects from low income housing to area parks and community centers (CCDC 1–12).

Downtown San Diego is also home to city, state, county, and federal government operations. The United States Navy and other military bases of operations, the United States Marshals Service, Immigration and Customs Enforcement, the Department of Homeland Security, a multi-story federal correctional facility, United States Postal facilities, United States Coast Guard Group San Diego, State of California administration buildings, San Diego County Administration buildings, county municipal and family courthouses, San Diego Police Headquarters, San Diego Fire Department Headquarters, San Diego Harbor Police, San Diego International Airport, San Diego City Hall, San Diego City administration buildings, libraries, transit operation headquarters including railroad and marine ports, colleges, a high school, and a senior center are all located in the 2.4 mile downtown San Diego. Over 75,000 miles of critical underground fiber optic cabling connects most of these organizations to the outside world (CCDC 1–12).

Given the above characteristics, San Diego qualifies as a prime potential target for any individual or group wishing to unleash a terrorist attack on a major U.S. city. Therefore, a detailed study of the city's information and communications technology (ICT) infrastructure will be very helpful in identifying vulnerabilities and proposing potential solutions to those vulnerabilities. It is the objective of this thesis to address these critical infrastructure issues.

RESEARCH QUESTIONS

The specific research questions that this thesis will address are the following:

1. How vulnerable to a terrorist attack is the information and communications technology (ICT) infrastructure in downtown San Diego?

3

2. How reliable are government and private institution connections to the existing ICT infrastructure?

3. Do the City of San Diego's policies for deploying fiber optic cabling between the city and the outside world ensure that fiber optic location information is shared among the private cable providers, thus protecting downtown's fiber optic network redundancy as a whole?

IMPORTANCE OF THE RESEARCH

The importance of this research is underscored by a closer look at the information architecture of downtown San Diego. Ideally, this examination further highlights the significance of the city's vulnerability to a terrorist attack. The fiber optic network vulnerability assessment and hypotheses developed in this thesis could potentially be applied to other geographic locations, or even at a national fiber optic infrastructure level.

Business and Government rely on information infrastructure to interconnect their critical communications and information systems to outside vendors, corporate headquarters, and to the public. Simple telephone communications, Voice over Internet Protocol (VoIP), video, cable television, Internet, multimedia conferencing, and even base stations for wireless cellular phone and Internet communication use underground fiber optic communication cable to interconnect their diverse locations. This critical fiber optic cabling runs through downtown San Diego public streets, sewer systems, and public right-of-ways, to provide connections from building wiring closets or point of presence (POPs),[1] to upstream connection providers. Using "9/11" as a paradigm, terrorists could plausibly cause the

---

[1] A point of presence is the location in a building which contains a collection of telecommunications equipment, usually modems, digital leased lines, and multi-protocol routers. This location is usually the first location that a communications provider's fiber optic lines would enter into an organization's building.

4

maximum amount of economic damage by targeting San Diego's ICT infrastructure. A report by the Conference Board of Canada provides the following assessment on the economic impact of the September 11, 2001 terrorist attack:

> The initial impact of the [September 11, 2001] terrorist attack on the U.S. economy will be very negative. An estimate of the damage to buildings and physical infrastructure in Lower Manhattan has already topped $20 billion. Economic activity was also severely disrupted. The nation's civilian air system and financial markets were largely shut down during the week of the attack and are only now being re-established. Job losses in the tens of thousands are already being announced by the nation's major airlines, as they significantly scale back flights as potential travelers stay home. The rest of the travel industry, including hotels, restaurants and convention centers are also in trouble. Note that much of service industry lost output will never be recovered, unlike the case of durable goods sales, which can simply be deferred. Naturally, the economic disruption goes well beyond these industries. The entire economy came to a virtual standstill in the few days after the attack as Americans watched their televisions to follow the rapid developments. (The Conference Board of Canada 1)

Given the above assessment, it is obvious that economic terrorism cannot be overlooked as a plausible means for terrorist attacks. Destroying or disabling downtown's network of fiber optic cabling could essentially disconnect downtown San Diego from the rest of the world. Without previously establishing multiple and diverse ICT connection providers, also known as "having redundancy," organizations in the downtown area would be unable to conduct essential business functions such as Internet access, e-mail, or telecommunications.

In the event of an attack or disaster, redundancy is an organization's ability to provide backup methods for external communications when an existing primary ICT provider fails. Many organization redundancy plans require different ICT connection providers to enter their building at opposite locations (often different ends of the building), in case of a physical disruption in service from an earthquake, vehicle collision into a POP, flood, other natural disasters, or even a well-planned terrorist attack.

It is hypothesized that most government and private organizations in the downtown area lack information on the location of ICT fiber optic lines, once these connections leave their respective buildings. In an interview on September 30, 2005, with Cynthia Queen, The City of San Diego Development Services Department, Queen stated that San Diego does not require fiber optic providers to communicate the location of existing fiber optic lines between providers, or even directly with the city. With respect to the continuity of enterprise business plans, it is hypothesized that these downtown organizations lack sufficient information of the full travel paths of each fiber optic line connected to their building. In this scenario, redundancy of ICT lines cannot be verified.

To completely verify that a redundant infrastructure exists, each street, intersection, waterway, structure, pipe, conduit, or any and all areas that each fiber optic line travels, must be known for each of the redundant paths. If two supposedly redundant ICT connection providers use the same path for their fiber optic lines in any one single location, these paths are no longer considered to be completely redundant of one another. Are ICT connection providers building their fiber optic infrastructure in close proximity to each other just to save money? Is a once separate and geographically diverse fiber optic network, becoming increasingly centric in downtown San Diego? Preliminary research shows that up to eight different ICT connection providers may travel through the same intersection and public right of way in downtown San Diego. It is important that these matters be addressed to ensure the economic survival of downtown San Diego during a natural or terrorist disaster.

RESEARCH METHODOLOGY

The potentially catastrophic problems and research questions posed in this thesis will be explored by conducting a CARVER + Shock Threat Assessment for the Information and Communication Technology System Infrastructure of downtown San Diego, California. During this threat assessment, potential information and communications system terrorist targets will be evaluated, as well as possible methods for conducting a successful terrorist attack. Also, the location of fiber optic cabling and critical system structures in downtown San Diego will be detailed.

Another outcome of the thesis is the creation of an information resource and report for the Department of Homeland Security, the State of California, County and City of San Diego, private industry ICT providers, and local businesses. These entities will have enough information to properly and quickly address security concerns and facilitate the building of necessary redundancies into downtown San Diego's vital ICT infrastructure.

ORGANIZATION OF THE THESIS

The rest of the thesis is organized as follows. Chapter 2 will discuss the vulnerability and criticality of fiber optic networks; Chapter 3 will analyze the threat to downtown's ICT infrastructure utilizing a CARVER + Shock Threat Assessment model; Chapter 4 will present the conclusions of the study and will propose recommendations for securing downtown's ICT infrastructure.

CHAPTER 2

FIBER OPTIC NETWORKS

The emergence of optical networking as a standard for high speed information sharing have made top telecommunication companies Level (3), AT&T, Nortel Networks, and Cisco famous. Installation of millions of miles of fiber optic cabling and expensive optical networking hardware kept company stock prices high and new fiber optic installations plentiful, especially during the late 1990s "Dot-Com boom" (Wikipedia, "Dark Fiber").

During this boom, telecommunication carriers were expected to spend billions of dollars every year to upgrade and to expand their existing fiber optic networks. Unfortunately, this expectation remained unreachable when the telecommunications market came to a screeching halt in 2001 (*Grid Today*, "How Does Optical Networking"). For example, the United States, the largest ICT spending nation, increased its spending only one percent between 2000 and 2001, compared with a fifteen percent spending increase in China (World Information and Technology Services Alliance, par. 1).

However, during a market turnaround in 2003, U.S. Companies again began spending upwards of $28.9 billion on information and communication technology equipment (U.S. Census Bureau x, fig. 2). Obviously, ICT infrastructure is critical to run daily business transactions, manage employee information, and transmit confidential or proprietary company data. High capacity fiber optic connections are crucial for powering daily business commerce and communication, from e-commerce to e-mail, and to sensitive government transactions. Without a concrete and highly redundant nationwide fiber optic network, it would be impossible to sustain daily business in the United States.

Telecommunication carriers rely on fiber optic connections much more than traditional wire-based architecture. Telecommunication providers build and maintain phone and Internet access, which powers daily business commerce. A New Yankee Group survey found that seventy-three percent of wire line service providers and thirty-one percent of wireless operators either have implemented or were testing packet telephony, using fiber optics instead of traditional copper wiring (Federal Communications Commission 5). An analysis of how telecommunication companies design, build and deploy fiber optic networks will help in understanding the critical vulnerabilities addressed in this thesis.

FIBER OPTIC CARRIER NETWORK DESIGN, CONSTRUCTION, AND RECORDS

In a telephone interview with Jim Hayes, President of The Fiber Optic Association, Inc. (a fiber optic industry educational trade group), Hayes claimed that he does not know of any specific national policy in existence for underground fiber optic construction in metropolitan areas. Most standards and policies are at the local level, thus enforced by the local municipal government. In fact, fiber optic carriers in the City of San Diego are not required to store location information for existing fiber optic infrastructure owned by other fiber optic carriers in the same area. In essence, multiple fiber optic lines could easily be just feet from one another without carriers ever knowing about this critical vulnerability. The question is whether this lack of information sharing seriously affects the fiber optic network's security as a whole. In addition, does the security of private carrier fiber optic lines become a public vulnerability with increasing government

9

use of private fiber optic lines for information transmission? An analysis of the City of San Diego's policies and procedures is needed to understand fiber optic line installations in pubic streets and right-of-ways, and how fiber optic installations are performed.

FIBER OPTIC INSTALLATION POLICIES AND PROCEDURES

The City of San Diego does not have a specific classification for fiber optic construction, but instead classifies fiber optics as "dry utilities." Dry utilities are all public utilities including but not limited to gas, electric, cable, telephones, fiber optic, traffic signals, street lights, and television. It does not include wet utilities such as reclaimed water, sewers, or storm drains (City of San Diego, "San Diego Municipal Code" Sept. 2003, 2). Dry utilities are installed in public right-of-ways located throughout the city (see fig. 1). Permits for dry utility installations are documented in the form of "D" sheets (see Appendix A) and are stored for public viewing in the City of San Diego Maps and Records section (Queen personal interview). The policies for dry utilities are listed in Appendix B.

Due to the sensitive nature of this figure,

it has been suppressed for general publication.

Figure 1. Selected view of public right-of-ways in a portion of downtown San Diego, California.
Due to the sensitive nature of this figure,
the source has been suppressed for general publication.

As previously mentioned, fiber optic cabling typically runs along public right-of-ways. The majority of right-of-ways in the City of San Diego are selected areas beneath paved public streets. According to Hayes, a right-of-way can be anywhere from a few feet to fifteen or twenty feet across. Public right-of-ways are streets, avenues, highways, lanes, and alleys that have been accepted according to law, or which have been in common and undisputed use by the public at large. Any private organization, with approval from the City of San Diego, may apply for a public right-of-way permit to utilize a public right-of-way. Private organizations' permits, however, are granted on a temporary basis and can be rescinded at any time by the City of San Diego. Public Utilities that are recognized by the California Public Utility Commission are granted access to the public right-of-way without having to apply for this permit or without having to pay a fee. These specific permits are non-revocable by the City of San Diego (Queen personal interview).

Once appropriate plans are approved and city permits are granted, a fiber optic installation can begin. The fiber optic construction company is charged with locating existing lines in the public right-of-way (r/w) around the construction site. Cynthia Queen, the supervising Public Information Officer for the City of San Diego Development Services Department, stated that it is "the applicant's responsibility to make sure they know about all existing facilities in the r/w. [Applicants] usually hire a professional engineer who spends a lot of time and effort in researching [The City of San Diego's] maps and records, and other resources to find out what has been constructed in the r/w" (Queen personal interview).

City policy and state law require a professional engineer to use a service such as the Underground Service Alert of Southern California, also known as DigAlert. DigAlert will contact utility companies that may have utility lines around the dig site. The utility companies have two days to mark

12

their existing lines before construction can begin. If a company that is building new facilities in the public right-of-way does not contact the DigAlert service, they can be charged with a misdemeanor, not to mention incurring liability for any financial loss sustained by utility carriers in the event of a line break.

The City of San Diego does not maintain a centralized, detailed, updated, or shared map of current fiber optic lines. In 1999, a project between the Center City Development Corporation (CCDC) and the City of San Diego Planning Department named "Bandwidth Bay" created an interactive map and GIS (Global Information System) layer showing the locations of fiber optic lines in downtown San Diego. The project took hard copy maps provided upon request from fiber optic carriers in downtown San Diego, and turned them into GIS shape files for use in GIS software and interactively at http://www.bandwidthbay.org. The project's original intention was to attract large companies to the San Diego area, highlighting the highly technological environment in downtown San Diego and the abundance of Internet bandwidth. The Bandwidth Bay fiber optic maps show which downtown streets are fiber optic routes, but does not show the exact location within the public right-of-way. The project has not been updated since late 2000 (Downtown San Diego Partnership, par. 6).

Fiber optic networks are crucial to government and private institutions alike. The rate at which information travels can affect critical decision making processes and mean the difference of thousands of dollars. Within the last decade, advancements in technology have led to the replacement of older copper-based cabling and relatively lower transmissions speeds with more advanced fiber optic cabling capable of unprecedented high-speed transmissions.

THE VULNERABILITY OF FIBER OPTIC NETWORKS

Fiber optic networks offer many advantages over older copper-based approaches, including less expensive long distance cable runs, ease of maintenance, and increased capacity. Fiber optic cabling even weighs less and is also smaller in size than copper cables. Since fiber optics use light to transmit signals, fiber is not subject to many common interferences such as radio or electromagnetic fields. The full capacity of fiber optic transmission speeds has yet to be realized (Macy 5).

As fiber optic technology progresses, faster speeds and longer transmission distances without the need for extra repeaters is continually developing and evolving. Fiber carriers have more interest than ever in developing new technologies to lower costs and increase their service offerings. There are, however, two highly troublesome trends. First, as fiber optic carriers compete for customers, they continually must lower prices while expanding their services. Delivering lower prices means that there must be a substantial company-wide increase in revenue or a decrease in costs. It is hypothesized that fiber optic companies are finding an easy but extremely critical area to start cost cutting: their network redundancy infrastructure. Second, as the transmission capacity for fiber optic cabling increases, the presence of a multitude of disparate network paths is decreasing. It is less expensive for carriers to run one fiber optic line and one trench than it is to run twenty lines over geographically diverse paths to equal the same capacity. This self-created vulnerability is known as network aggregation.

Network Aggregation

The general trend of fiber optic carriers to build fiber optic lines closer to one another and with less geographic diversity is known as network aggregation. A paper published by The National Academies, "Computer Science and Telecommunications Board," revealed that this network aggregation of fiber optic lines was recognized as a problem as early as 1989:

> The competitive environment of the past decade has caused a proliferation of networks and network vendors. One would think this proliferation should decrease vulnerability because of the added redundancy provided by multiple networks. However, in practice, there is actually less than meets the eye. Some of these networks traverse the same geographic rights-of-way and are thus vulnerable to the same physical attacks. Further, competitive factors and the large array of technical alternatives have increased incompatibilities in public and private networks, so that it will be difficult to use the surviving assets in one network to back up those of another. (14)

Michael A. Caloyannides, PhD, of Dartmouth College, published a whitepaper entitled *Potentially Catastrophic Vulnerabilities of the Internet and Proposed Remedies*, where he identified the real threat to networks. According to Caloyannides, it is not the temporary software-based attacks that are problematic, but rather, the simple and devastating attacks to key physical infrastructure. And physical infrastructure is continually agglomerating to smaller geographic areas:

> The commercialized Internet of today is driven by market economics and not by security considerations. Cost-effectiveness has dictated the shift towards major nodes and paths ('backbones') whose workload cannot be taken over by ancillary smaller communications paths and nodes. If such key paths and nodes are rendered inoperative, the Internet will at best slow down to levels that make it essentially useless; users that require fast speed (such as IP telephony, online business, etc.) will stop altogether . . . Denial of Service and virus attacks are a temporary nuisance, while assaults to key and publicly known critical nodes would be devastating to entire segments of the network. (Caloyannides 13)

Caloyannides also found that key infrastructure maps of Internet backbones are publicly available, and accessible worldwide, " . . . from many sources, such as various companies' marketing departments" (Caloyannides 13).

Sean P. Gorman, an Associate Research Professor at George Mason University and author of the doctoral thesis *Networks, Complexity, and Security: The Role of Public Policy in Critical Infrastructure Protection* argues that, "private efficiencies can result in public vulnerabilities" (162). Gorman further explains that high-order economic activities tend to concentrate in large metropolitan areas, while low-order economic activities have increasingly been exported to outlying regions, away from main command-and-control urban centers. Urban centers have become increasingly important to business operations because of the vast low-cost labor pool and the aggregation of technology infrastructure. Gorman applied this theory to extend to "telecommunications and IT: the higher the telecommunications and information technology density, the more productive the agglomeration" (138). Several researchers agree that "backbone networks disproportionately agglomerate in the largest metropolitan areas" (39).

In order to link these low and high-order economic activity locations, fiber optic lines must pass information to and from downtown San Diego and outlying regions. Chapter 3 will concentrate on analyzing the specific vulnerabilities of these fiber optic transmission lines, which transmit critical information to and from downtown San Diego. Not only has an increase in technology contributed to this serious problem, but also the City of San Diego itself has contributed to the reduction of the geographic distance between fiber optic carriers.

Private Sector Efficiencies

The City of San Diego is keenly aware of industry's momentum to reduce costs. During the recent construction of the ballpark area in downtown San Diego, the City of San Diego Engineering and Capital Projects Department released a memo to all dry utility companies (see Appendix C) which reads, in part:

> the City has become aware of an ongoing effort by private entities, to coordinate planning and construction of a joint trench to share in the costs and expenses of design and construction of new facilities in the streets of East Village. Because there is a limited amount of available space in the streets in the East Village area, the City is encouraging every effort to plan ahead and to coordinate future construction. (par. 3)

Furthermore, "participation in this *joint trench* may provide efficiency and realization of *cost savings* for those involved" (par. 3).

San Diego-based energy supplier, Sempra Energy, has potentially added another catastrophic vulnerability to the list of private sector efficiencies. In late 2003, The California Public Utilities Commission approved Sempra Fiber Link's "last-mile" telecommunications solution, to install fiber optic cabling in existing natural gas pipeline. The "FIG" natural gas line installation method saves money, avoids trenching, and allows gas companies to supply additional services such as broadband Internet, to its gas customers. If the San Diego Gas and Electric fiber optic lines in downtown San Diego traverse through natural gas pipelines, this would encase critical communication links in a potentially explosive gas. These same fiber optic lines could help terrorists locate natural gas lines. Sempra's Fiber Links subsidiary dissolved in late 2005 (Sempra Energy, "California Utility Agency").

The high cost, effort, and the time of permitting and constructing new rights-of-way is encouraging fiber optic carriers to concentrate their geographic routes (The National Academies 24). The cost to map locations of existing utility lines in the right-of-way, and of designing new geographically diverse routes, outweigh the benefits, in the view of fiber optic providers. This creates reliance on specific geographic locations for fiber optic network survivability and thus creates a single point of failure.

WHY ATTACK FIBER OPTIC INFRASTRUCTURE?

Cyberterrorism[2] is a popular word used in both mainstream media and in advertising. From virus attacks to email phishing scams, headlines detail the possibility of shutting down the Internet through cyberterrorism. Is cyberterrorism or cyberhooliganism the real threat to networks and ICT infrastructure?

Bruce Schneier, Chief Technical Officer of Counterpane Security, makes a distinction between the term "cyberterrorism" and "cyberhooliganism." To date, there has never been a substantiated act of cyberterrorism against the United States—it is unclear whether or not Schneier implies that all computer attacks reported thus far in the United States fall under "cyberhooliganism." In Maroochy Shire, Australia in April 2000, a disgruntled consultant compromised a waste control management system and released millions of gallons of untreated waste into the town. What the media did not reveal was that it took this former company insider 46 attempts to release the waste. He also had previously installed the proprietary water control management software on his laptop before leaving the company.

In 2004 alone, deliberate sabotage to both a main (underground) and backup (aboveground) fiber optic network cable resulted in 250,000 Bell Canada customers without phone service, Internet, and essential 911 service

---

[2] Cyberterrorism: a criminal act perpetrated through computers resulting in violence, death and/or destruction, and creating terror for the purpose of coercing a government to change its policies.

and emergency communications. The individuals involved knew the exact locations of the cables and the importance to Bell Canada's network (Converge! Media Ventures, Inc. par 1). Yet Bell Canada considered their fiber optic cables well protected because, "the underground cables were buried in enclosures several feet underground. They were further protected by wooden covers, gravel and a large, heavy boulder. The aboveground cable is located at the top of a 20-foot pole and would also take significant time and knowledge to access" (Bell Canada Enterprises, par. 5).

A number of authors remain steadfast that the threat from cyberterrorism is only a myth:

> There is no such thing as cyberterrorism—no instance of anyone ever having been killed by a terrorist (or anyone else) using a computer. Nor is there compelling evidence that al Qaeda or any other terrorist organization has resorted to computers for any sort of serious destructive activity. (Green, par. 6)

Terrorists will follow the path of least resistance. Physical attacks, such as conventional explosives, offer a cheap and easy alternative to intricate "cyberwarfare" (Berinato, "The Truth About").

Since fiber optic cabling runs over public right-of-ways in downtown San Diego, fiber optic cables have only the protection of concrete and asphalt above their underground entrenchment. Furthermore, most fiber optics require a repeater or amplifier to maintain signal strength at some point during the fiber optic cable run. These signal strengthening devices refresh signals that travel large distances. Major Gerald Hust, United States Air Force, argues that these switching facilities are "the most critical elements in a telecommunications system. They are often highly automated, unmanned [or lightly manned], and remotely monitored" (8). Because of the risks involved in revealing specific and proprietary information about each carrier's fiber optic network, this thesis will not address the vulnerability of specific switching stations in downtown San Diego.

Wireless Extended Fiber Optic Networks

An increasingly troublesome pattern of ideology has been discovered among downtown San Diego business owners. Many businesses deploy wireless-based ICT infrastructure to connect to the Internet, and to utilize critical outside business communication services such as VoIP, or wireless cellular phones. The belief that these services are not linked to downtown San Diego's fiber optic infrastructure is flawed. The National Academies Press identified that a number of terrestrial microwave and radio routes track closely to rights-of-way for both coaxial and fiber optic cable routes. Fiber optics were also hypothesized to replace satellite voice services as early as 1989 (24). Furthermore, both "wired and wireless infrastructures are increasingly tethered to a fiber optic (transit) backbone for connectivity into the domestic and global network" (Gorman 61). A publicly accessible map of wireless cellular telephone towers in downtown San Diego is listed in Appendix C. The locations of these cellular tower installations are near perfectly correlated to routes used by underground fiber optic cabling.

Fiber optic cabling is replacing both microwave and satellite for backbone telecommunication services. Satellite and microwave transmission systems cannot compete economically or practically with both undersea or terrestrial fiber optic systems (Gorman 61).

The Wireless Consumers Alliance (WCA), a non-profit cell phone consumer group, has identified this wireless to wired vulnerability as a critical ICT infrastructure issue. The WCA admits that "in an emergency situation, there will be some loss of [wireless network] capacity. This situation is often exacerbated by the fact that fire and safety personnel use wireless systems as a back up to their own wired facilities" (par. 3).

THE EFFECT OF CONVENTIONAL EXPLOSIVES ON UNDERGROUND FIBER OPTIC NETWORKS

A bomb equal to 4,000 pounds of trinitrotoluene (TNT) detonated alongside the Oklahoma City Murrah Federal Building in 1995. The explosive, located in a large rental truck on a chassis 4.5 feet above the street, created a blast area 28 feet wide, and penetrated 6.8 feet underground—easily destroying the 11 inches of asphalt and 7 inches of concrete which comprised the city street, and protected the fiber optic infrastructure underneath (see fig. 2).

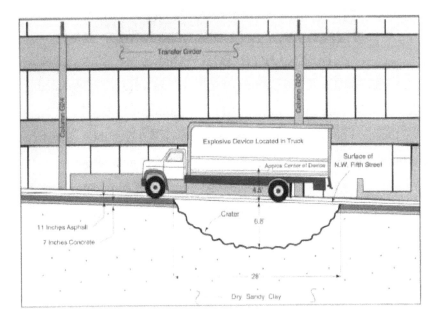

Figure 2. The Oklahoma Federal Building bombing demonstrated the impact a conventional weapon can have on the concrete and asphalt of city streets. City streets are some of the only protection for underground fiber optic cables.
Source: W. Gene Corley, "Applicability of Seismic Design in Mitigating Progressive Collapse" July 10, 2002. Nov. 25, 2005, <http://www.nibs.org/ MMC/ProgCollapse%20presentations/Corley's%20revised%20paper.pdf> p. 10, fig. 3.

There is limited research available on the effects of conventional explosives on underground fiber optic cabling. Underground cabling is protected by asphalt, concrete, and soil that encase it from above. The City of San Diego does not require any standard protective measures for fiber optic cables, such as protective conduit, tamper alarms, or monitoring devices. Utilities such as electric lines or gas mains have specific policies for safety and security.

The direct effects to underground fiber optic cabling buried in San Diego City streets from conventional explosive detonation are generated from the following characteristic effects of an explosion:

1. Blast— the high overpressure creating shock such as that found in a fuel air explosive weapon.
2. Penetration—a bomb or fragments of jagged steel produced from the bomb casing exploding or special devices inside the casing that penetrate the target breaking the system or its subsystems.
3. Crater—a violent earth shock breaking up smooth surfaces or damaging them so that the surface becomes unusable. This is the result of penetration and blast.
4. Fire—fire damage caused by weapons and fires fueled by target material itself with radiant heat igniting combustibles to melt and damage components. (Hust 47)

San Diego streets have varying depths and consistencies of concrete and asphalt depending on several factors, from the amount of traffic to soil stability. An excerpt from the City of San Diego Pavement Design Standards, Schedule J, is displayed in Table 1.

Using the Murrah Federal Building bombing as a contrast, Table 2 illustrates a pavement supposition, showing the amount and consistency of varying street compositions. City designers use soil quality (R-value), and traffic indices, to set minimum depths for asphalt/concrete (A/C), and cement-treated base (CTB). The street classification ranges from a typical commercial street (Collector) to a major 6 lane road. Street designers can use any of the three types of street designs: (1) Flexible Pavement Section, (2) Rigid Pavement Section, and (3) Full Depth AC (asphalt/concrete) depending on what the design and conditions warrant.

The Alfred Murrah Building column (4) shows the depth of both AC and either CTB or PCC in the street, adjacent to the Murrah federal building, where the 4,800 lb. ammonium nitrate fuel oil bomb created a 6.8 foot crater in the city street. The City of San Diego does not use vulnerability or criticality assessments of underground fiber optics as a method for determining the required depth of asphalt or concrete for downtown city streets.

Finally, Table 3 shows a comparison of the Flexible Pavement Section, Rigid Pavement Section, and Full Depth AC, to the asphalt and concrete base at the Murrah Federal Building bomb site. It is disturbing to note that of the city's thirty-six possible street asphalt and concrete construction depths, only the eight highlighted in blue have thicknesses exceeding the asphalt and concrete used at the Murrah Federal Building bomb site.

The Federal Emergency Management Agency (FEMA) in its publication "A How-To Guide to Mitigate Potential Terrorist Attacks Against Buildings" provided information to assist private industry in identifying potential destructive capabilities to buildings (see fig. 3).

Table 1. San Diego City Street Asphalt and Concrete Standards

Street Classification	"R" Value[a]	Flexible Pavement Section		Rigid Pavement Section		Full Depth AC	Alfred P. Murrah Building	
		AC (in)[a]	CTB (in)[a]	PCC (in)[a]	CTB (in)	(in)	AC (in)	CTB/PCC? (in)
Collector (Commercial/Industrial)	0-9.9	5	18.5	8.5	6	15.5	11	7
	10-19.9	4.5	16.5	8.5	5	14	11	7
	20-29.9	4	15	8.5	0	13	11	7
	30-39.9	3.5	13	8.5	0	12	11	7
	40-49.9	3	11	8	0	11	11	7
	50 or Greater	3	8.5	7.5	0	9	11	7
Major 6-Lane	0-9.9	6.5	21	9	6	18.5	11	7
	10-19.9	6	18.5	9	6	16.5	11	7
	20-29.9	5	17	8.5	5	15.5	11	7
	30-39.9	4.5	15	9	0	14	11	7
	40-49.9	4	12.5	8.5	0	12.5	11	7
	50 or Greater	3.5	10.5	8	0	11	11	7

Source: City of San Diego. "Standard Drawings," July 1, 2004. Nov. 25, 2005, <http://www.sandiego.gov/engineering-cip/pdf/pdf/csdsd2003.pdf> SDG-113, sheet 3.

[a] AC = asphalt/concrete, in = inches, CTB = cement treated base, PCC = Portland cement concrete, "R" Value = soil quality.

Note. Interpretation courtesy of Josh Butler, Civil Engineer, Boise, ID.

Table 2. City of San Diego Pavement Supposition

Street Classification	Max ADT	Max Traffic Index	"R" Value [a]	Standard Sections AC (in)[a]	Standard Sections CTB (in)[a]	Concrete M.O.R. 600 MIN PCC (in)[a]	Concrete M.O.R. 600 MIN CTB (in)	Full Depth AC (in)
Cul-de-sac	200	5.0		3.0	5.0	6.0	---	4.5
Local (L.V.R.)	700	5.5		3.0	5.0	6.5	---	5.0
Local (Res.)	1200	6.0		3.0	5.0	6.5	---	5.5
Local (Res.)	2200	6.5		3.0	5.0	6.5	---	6.0
Local (Ind.)	2000	8.5		3.0	7.5	7.5	---	8.5
Collector (Res.)	3500	7.0		3.0	5.0	7.0	---	6.5
Collector (Res.)	5000	7.5	50.0	3.0	5.5	7.0	---	7.5
Collector (Comm./Ind.)	5000	9.5	or	3.0	8.5	7.5	---	9.0
Collector (No Frt.)	7500	8.0	greater	3.0	6.5	7.0	---	8.0
Collector	15000	9.0		3.0	7.5	7.5	---	8.5
Major (4-lane)	30000	10.5		3.0	10.0	8.0	---	10.5
Major (6-lane)	40000	11.0		3.5	10.5	8.0	---	11.0
Primary Arterial	50000	11.5		3.5	11.5	8.0	---	11.5
Expressway	60000	12.0		3.5	11.5	8.5	---	12.0
Expressway	80000	12.5		4.0	12.0	8.5	---	12.5
Expressway	100000	13.0		4.0	12.5	9.0	---	13.0

Source: City of San Diego. "Standard Drawings," July 1, 2004. Nov. 25, 2005, <http://www.sandiego.gov/ engineering-cip/pdf/csdsd2003.pdf> SDG-113, sheet 2.

[a] AC = asphalt/concrete, in = inches, CTB = cement treated base, PCC = Portland cement concrete, "R" Value = soil quality.

Table 3. Total Thickness Comparison of Several Types of City of San Diego Streets, Compared to the Street Damaged in the Murrah Federal Building Bombing

Total Thickness Comparison (in.)			
Flexible Pavement Section	Rigid Pavement Section	Full Depth AC	Alfred P. Murrah Building
23.5	14.5	15.5	18
21	13.5	14	18
19	8.5	13	18
16.5	8.5	12	18
14	8	11	18
11.5	7.5	9	18
27.5	15	18.5	18
24.5	15	16.5	18
22	13.5	15.5	18
19.5	9	14	18
16.5	8.5	12.5	18
14	8	11	18

Source: City of San Diego. "Standard Drawings," July 1, 2004. Nov. 25, 2005, <http://www.sandiego.gov/engineering-cip/pdf/csdsd2003.pdf> SDG-113, sheet 3.

FEMA does not account for effects to property, such as ICT infrastructure or city streets in their publication. FEMA identifies a large semi truck as having the ability to yield an explosion equal to 100,000 lbs of TNT. A large vehicle, carrying a large number of explosives and strategically placed over an underground fiber optic installation, may be an effective weapon against ICT infrastructure—as exemplified during the bombing of the Murrah Federal Building.

If an explosive equal to 4,000 pounds of TNT can create a 6.8 foot crater in eleven inches of asphalt, and seven inches of concrete at the Murrah Federal Building, the effect of a 100,000 pound explosion would most definitely exceed the protection afforded to underground fiber optic infrastructure by most current city street designs.

Only eight of thirty-six possible street thicknesses exceed the Murrah Federal Building street thickness, which was still penetrated a total of 6.8 feet by an explosive equal only to 4,000 lbs of TNT.

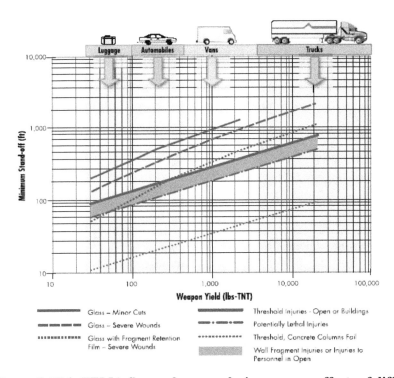

Figure 3. This FEMA figure shows explosive weapon effects of different weapon yields (in lbs. of TNT) from different stand-off distances. It also estimated the amount of explosive power that can be contained in different styles of delivery vehicles such as luggage, or a large semi-truck.
Source: Federal Emergency Management Agency. "Risk Management Series: Risk Assessment—A How-To Guide to Mitigate Potential Terrorist Attacks Against Buildings," *FEMA* 452 Jan. 2005. Nov. 25, 2005, <http://www.fema.gov/pdf/fima/452/fema452.pdf> p. 6, fig. 1-4.

SUMMARY

Fiber optic networks are the backbone of San Diego's ICT infrastructure. Government and private industry use fiber optics for everything from daily business email to national security matters.

The City of San Diego does not have specific policies for fiber optic network installations in city-owned streets. Fiber optic carriers do not have enough information to pinpoint the exact locations of fiber optic installations by other carriers, in order to plan for diversity in their fiber optic routes—creating agglomeration. This lack of concrete policy and location information for private fiber optic carriers creates a public vulnerability to business and government in downtown San Diego.

Additionally, wireless cellular phone towers, mobile Internet access, and many microwave radio installations are closely tied to land-based fiber optic routes. Many local business owners in downtown San Diego do not understand this direct link between wired and wireless services. Conventional weapons are the easiest form of attack for underground fiber optic infrastructure. The bombing of the Murrah Federal Building, with a conventional explosive equal to 4,000 lbs of TNT, demonstrated the effect conventional weapons have on the asphalt and concrete of city streets— easily tearing through fiber optic lines buried just under the surface.

Chapter 3 will explain the CARVER + Shock threat assessment process, and culminate in a full threat assessment of the information and communication technology in downtown San Diego.

CHAPTER 3

CARVER + SHOCK THREAT ASSESSMENT FOR INFORMATION AND COMMUNICATION TECHNOLOGY SYSTEM INFRASTRUCTURE OF DOWNTOWN SAN DIEGO, CALIFORNIA

DEFINITION

For this CARVER + Shock threat assessment of downtown San Diego, California, the area defined as "Downtown" San Diego will closely resemble the designation of "Downtown" by the San Diego Center City Development Corporation (CCDC). This is an approximate 1,500 acre land parcel bordered on the north by Laurel Street, to the east by Interstate-5, to the south by Commercial Street, and to the west by the Pacific Ocean (including downtown San Diego's ports and piers).

The fiber optic line location information is taken from the CCDC/City of San Diego Bandwidth Bay project. The fiber optic data is current as of 1999 and is specific only to the streets in which the fiber optic lines traverse. Fiber optic lines in downtown San Diego travel down public rights-of-way, which are mapped through data from San Diego Geographical Information Source (SanGIS), and Southern California DigAlert.

All information used in compiling this threat assessment is open source and obtained using publicly available media, either for a fee or free of charge through research or public information requests.

THREAT ASSESSMENT PURPOSE

The threat assessment described in this Chapter will focus on the vulnerability of the Information and Communication Technology (ICT) infrastructure in downtown San Diego from conventional explosives. Specifically, the threat assessment focuses on the heart of ICT infrastructure—underground fiber optic cabling that connects downtown commerce and government to the outside world. These fiber optic cable routes travel along public right-of-ways in downtown San Diego streets.

This threat assessment can be used to help both public and private industry planners increase redundancy and mitigate possible effects from natural disasters, fiber optic sabotage, and accidental utility breakage. It will not offer a threat assessment that accounts for chemical, biological, radiological, or nuclear devices. Analyzing the latter threats would require significantly more research and specific expertise into their effects on ICT infrastructure, specifically on underground fiber optic cabling.

Furthermore, the threat assessment will not focus on specific downtown businesses or addresses, but will focus on negative trends such as agglomeration, policy shortfalls, location of critical infrastructure, and information sharing inefficiencies that have been explored in Chapters 1 and 2 of this thesis.

PROCEDURE

In order to better understand and visually conceptualize the ICT infrastructure in downtown San Diego, ESRI® ArcMap™, a Global Information System (GIS) software application, was used to combine over 16 GIS overlays on top of one-foot color aerial photography of downtown San Diego. GIS Layers were included from the San Diego County Department of Planning and Land Use, City of San Diego Maps and Records Division, the

United States Census Bureau, United States Geological Survey—Terraserver, and the San Diego Geographical Information Source (SanGIS) (see fig. 4). Combining Bandwidth Bay fiber optic cabling street location information and SanGIS public right-of-way maps, fiber optic cabling can be mapped within feet of its actual location. With the help of a service such as DigAlert, the accuracy can be precise to within inches. It is highly unlikely that fiber optic providers have constructed additional fiber optic routes since the Bandwidth Bay project concluded in 1999 (with the exception of the Ballpark area).[3] This helps to validate the accuracy of the fiber optic location information for use in the year 2006.

The nature of Internet and fiber optic connections make it difficult to analyze the measurable effect during a fiber optic line break. Without analyzing network maps from individual fiber optic providers, it would be difficult to ascertain exactly how damage to fiber optic connections in one location of a fiber optic company's network could affect the entire network's connectivity in downtown San Diego. Many fiber optic companies deploy "self-healing" fiber optic network infrastructures, which allow rerouting of traffic using undamaged fiber optic routes. By analyzing system-wide vulnerabilities, such as an over or under saturation of ICT infrastructure in a certain geographical area, more serious ICT system-wide vulnerabilities can be discovered without having to ascertain the precise backup architecture of each provider's infrastructure.

---

[3] City of San Diego Planning employee who requested anonymity.

Due to the sensitive nature of this figure,

it has been suppressed for general publication.

Figure 4. Downtown San Diego 1-Ft Satellite Map and Fiber Optic GIS Layers from ESRI® ArcMap™. Downtown defined by orange border. Due to the sensitive nature of this figure, the source has been suppressed for general publication.

The nature of fiber optic routes in downtown San Diego can be analogized to a spider web connected to a house. If several inner strands of the spider's web were broken, the effects to the entire web would be minimal, while an attack on specific supporting strings of the web (vulnerable locations in the system), could prove catastrophic for the entire spider web. A terrorist would be more effective in striking these critical points because they are vulnerable concentration points of ICT network traffic reaching downtown organizations. A CARVER + Shock Threat Assessment will be conducted to evaluate threats to these critical ingress/egress points.

CARVER + SHOCK METHODOLOGY

CARVER + Shock was originally developed as a targeting analysis model by the United States Air Force to evaluate and prioritize significant military targets. Target analysis is used to evaluate potential targets to find the most effective attacks, using the least amount of effort or personnel (National Infrastructure Institute, par. 3).

Each letter in C-A-R-V-E-R represents a specific aspect of a prospective terrorist target. "C" stands for the criticality of the target, "A" for target accessibility, "R" for recoverability or recuperability, "V" for vulnerability, "E" for effect on the target, and "R" for recognizability factors. The CARVER + Shock model is unique because it adds a component of measurement (+ Shock—the combined measure of the psychological and collateral national economic impacts of a successful attack on the target) that accounts for national pride, time, and other factors once considered immeasurable.

CARVER + Shock is also different from many traditional target analysis models because its assessment criteria are easily adaptable to specific types of targets. For example, the United States Department of Agriculture, Food and Safety Inspection Service has customized assessment criteria in the CARVER + Shock model, in order to analyze specific targeting aspects for food-borne diseases and Agroterrorism. The effect on the food chain, and the security of food producers, are examples of just a few custom target criteria the Department of Agriculture has built into their food-specific CARVER + Shock threat assessment model. By customizing the criteria used to score CARVER aspects, the overall threat assessment is more in-tune to account for specific issues concerning threats to food production (Food and Drug Administration, slides 9–29).

In this thesis, a variation of the CARVER + Shock Threat Assessment model called the Information and Communication Technology infrastructure variation, has been created and customized to better analyze threats to ICT infrastructure targets. While the ICT variation of the CARVER + Shock model still uses the basic principles of the CARVER + Shock model, it can better account for target criteria specific to ICT infrastructure.

34

The results from using the CARVER + Shock model form a usable real-life threat assessment to "reverse engineer" an attack. A customized CARVER + Shock model using targeting criteria specific to ICT infrastructure is an excellent model to evaluate downtown San Diego's ICT infrastructure.

Each target criterion will be evaluated using a numerical scale with values between 1 and 10 as indicated in Tables 4–10.

Criticality or "C" in CARVER + Shock is the importance of a system, subsystem, complex, or component. A target is critical when its destruction or damage has a significant impact on the output of the targeted system, subsystem, or complex. Scoring metrics are found in Table 4.

Accessibility is the ease with which a target can be reached. A target is accessible when an attacker can reach the target to conduct the attack and egress the target undetected. Accessibility is the openness of the target to the threat. Scoring metrics are found in Table 5.

Table 4. Scoring Metrics of Criticality

Criteria	Scale
Loss of connectivity to 60% or more of information or communication technology (ICT) infrastructure OR 60% or more of critically connected enterprises or operations affected.	9 – 10
Loss of connectivity to 40-60% of ICT infrastructure OR 40-60% of critically connected enterprises or operations affected.	7 – 8
Loss of connectivity to 20-40% of ICT infrastructure OR 20-40% of critically connected enterprises or operations affected.	5 – 6
Loss of connectivity to 0-20% of ICT infrastructure OR 0-20% of critically connected enterprises or operations affected.	3 – 4
No Loss of connectivity to ICT infrastructure OR No critically connected enterprises or operations affected.	1 – 2

Table 5. Scoring Metrics of Accessibility

Criteria	Scale
Easily Accessible (e.g., ICT infrastructure is above ground with no perimeter security). No physical or human barriers or observation. Attacker has unlimited access to the target. Attack can be carried out without undue concern of detection. Multiple sources of information concerning the infrastructure and the target are easily available.	9 – 10
Accessible (e.g., ICT infrastructure is underground but has minimal perimeter security). Human observation and physical barriers limited. Attacker has limited time. Attack can be carried out but requires the use of stealth. Only limited specific information is available on the infrastructure and the target.	7 – 8
Partially Accessible (e.g. ICT infrastructure is underground but has limited perimeter security). Under periodic possible human observation. Some physical barriers may be present. Time limitations are significant. Only general, non-specific information is available on the infrastructure and the target.	5 – 6
Hardly Accessible (e.g., ICT infrastructure is underground and has moderate perimeter security). Human observation and physical barriers with an established means of detection. Access generally restricted to operators or authorized persons. Time limitations are extreme. Limited general information available on the infrastructure and the target.	3 – 4
Not Accessible. Physical barriers, alarms, and human observation. Defined means of intervention in place. Attacker can access target for less than 5 minutes. No useful publicly available information concerning the target.	1 – 2

Recoverability or recuperability is a measure of the time required to replace, repair, or bypass the destruction or damage inflicted to the target. CARVER Redundancy levels are typically measured using a time factor for recovery between one month and one year. In today's fast-paced, Internet-driven economy even one hour of time to recovery may be debilitating to an enterprise or operation. The concept of Infinite Parallel Redundancy[4] (MathPages, par. 1) is a better criterion for evaluating a target's recoverability because it uses a "mean time between failure" (MTBF) approach, rather than a set numerical time. "N" represents a single information or communication route (typically a single fiber optic route). "N+1" is a single information or communication route, plus an additional route. Other "N+" measurements are also measured similarly. Scoring metrics are found in Table 6.

Table 6. Scoring Metrics of Recuperability/Recoverability

Criteria	Scale
No redundant routes available OR three + days to re-establish connection for critical enterprises or operations.	9 – 10
N+1 redundancy OR one to three days to re-establish connection for critical enterprises or operations.	7 – 8
N+2 redundancy OR 24 hours to one day to re-establish connection for critical enterprises or operations.	5 – 6
N+3 redundancy OR 12 to 24 hours to re-establish connection for critical enterprises or operations.	3 – 4
N+4 redundancy OR <12 hours to re-establish connection for critical enterprises or operations.	1 – 2

---

[4] A common strategy for increasing the mean time between failure (MTBF) of a system is to add redundant paths in parallel. If each path has an exponential failure rate of q per hour, then its MTBF is $1/q$ hours, and the MTBF of two parallel paths is $3/(2q)$. In other words, the redundant path increases the MTBF by 50%.

Vulnerability is a measure of the ease of achieving the attacker's purpose once the target has been reached. Vulnerability is determined both by the characteristics of the target (e.g. nature and construction of the target, amount of damage required), and the characteristics of the surrounding environment (ability to work unobserved, time available to attack target). Scoring metrics are found in Table 7.

Effect is the percentage of daily infrastructure productivity affected by the attack. Scoring metrics are found in Table 8.

Recognizability is the degree to which a target can be recognized under varying conditions (e.g., weather or special events) without confusion from other targets or components. Factors that influence recognizability include:

- The size and complexity of the target
- The existence of distinctive target signatures

Scoring metrics are found in Table 9.

Table 7. Scoring Metrics of Vulnerability

Criteria	Scale
Threat can impose desired effect under any condition.	9 – 10
Threat can impose desired effect under ideal conditions.	7 – 8
Threat can impose 50% of desired effect under any condition.	5 – 6
Threat can impose 50% of desired effect under ideal conditions.	3 – 4
Threat cannot impose desired effect under any condition.	1 – 2

Table 8. Scoring Metrics of Effect

Criteria	Scale
Greater than 50% of system-wide ICT production impacted.	9 – 10
25-50% of system-wide ICT production impacted.	7 – 8
10-25% of system-wide ICT production impacted.	5 – 6
1-10% of system-wide ICT production impacted.	3 – 4
Less than 1% of system-wide ICT production impacted.	1 – 2

Table 9. Scoring Metrics of Recognizability

Criteria	Scale
The target is clearly recognizable and requires little or no training or preparation for recognition.	9 – 10
The target is easily recognizable and requires only a small amount of training or preparation for recognition.	7 – 8
The target is difficult to recognize or might be confused with other targets or target components and requires some training or preparation for recognition.	5 – 6
The target is difficult to recognize. It is easily confused with other targets or components and requires extensive training or preparation for recognition.	3 – 4
The target cannot be recognized under any condition, except by experts.	1 – 2

Shock is the final attribute considered in the CARVER + Shock methodology. Shock is the combined measure of the psychological and collateral national economic impacts of a successful attack on the target. Shock is usually considered on a national level. The psychological impact will be increased if there are a large number of deaths or the target has historical, cultural, religious or other symbolic significance. Scoring metrics are found in Table 10.

Table 10. Scoring Metrics of Shock

Criteria	Scale
Target has major historical, cultural, religious, or other symbolic importance. Loss of over 10,000 lives. Major impact on sensitive subpopulations, e.g., children or elderly. National economic impact more than 100 billion dollars.	9 – 10
Target has high historical, cultural, religious, or other symbolic importance. Loss of between 1,000 and 10,000 lives. Significant impact on sensitive subpopulations, e.g., children or elderly. National economic impact between ten and 100 billion dollars.	7 – 8
Target has moderate historical, cultural, religious, or other symbolic importance. Loss of life between 100 and 1,000. Moderate impact on sensitive subpopulations, e.g., children or elderly. National economic impact between one and ten billion dollars.	5 – 6
Target has little historical, cultural, religious, or other symbolic importance. Loss of life less than 100. Small impact on sensitive subpopulations, e.g., children or elderly. National economic impact between 100 million and one billion dollars.	3 – 4
Target has no historical, cultural, religious, or other symbolic importance. Loss of life less than 10. No impact on sensitive subpopulations, e.g., children or elderly. National economic impact less than 100 million dollars.	1 – 2

In order to assemble and complete a single CARVER + Shock Threat Assessment the following steps will be completed in this thesis:

1. Identify vulnerabilities and make observations.
2. Assess risks.
3. Prioritize vulnerabilities.
4. Brainstorm countermeasures and make appropriate recommendations.

(This will be analyzed further in Chapter 4: Conclusions and Recommendations).

In order to prioritize vulnerable points of failure in downtown's fiber optic network, a CARVER + Shock Threat Matrix will be constructed to assign numerical values to targets based on their overall threat score, including individual target selection factors (C-A-R-V-E-R + Shock). Each of the factors in CARVER + Shock are equally weighted for each target, in order to find an overall target threat score.

CARVER + SHOCK THREAT MATRIX

The CARVER + Shock threat matrix is a decision tool for rating the relative desirability (from the attacker's point of view) of potential targets and for wisely allocating attack resources (or-conversely implementing countermeasures, from the point of view of the potential victim). The threat matrix lists selected ICT infrastructure locations by their overall target score. A numerical rating system (one to ten) will rank the CARVER + Shock factors for each potential target. A score of ten indicates a highly desirable rating (from the attacker's point of view), and a score of one reflects a highly undesirable rating. An overall score is also tabulated for each target. The CARVER + Shock Threat Assessment Matrix provides a more tactical assessment of the probability of attack within a particular threat factor. Higher numbers indicate a greater likelihood for attack (Clark 78).

THREAT ASSESSMENT

The entire CARVER + Shock Threat Assessment workbook and threat matrix are located in Appendix E. However, due to the sensitive nature of this appendix, it has been suppressed for general publication.

SUMMARY

The CARVER + Shock threat assessment model was developed as a tool to help evaluate and prioritize threats to potential terrorist targets. Data used in compiling this threat assessment is all open source and publicly available.

CARVER + Shock is a strong threat assessment model because it allows the person conducting a threat assessment to exercise judgment within each target criterion, instead of relying merely on computer calculations or standard threat models.

The custom CARVER + Shock ICT infrastructure threat assessment variation, is more specifically tailored to ICT infrastructure threat targets, while at the same time relying on the basic principles of CARVER + Shock.

Results from the threat assessment are assembled into a threat matrix. The threat matrix prioritizes targets based on their overall threat score. This matrix can be used by planners, to better prioritize security and infrastructure protection resources.

Chapter 4 will address conclusions found in this thesis, and offer possible recommendations to address vulnerabilities.

CHAPTER 4

RECOMMENDATIONS AND
CONCLUSIONS

In this thesis the threat assessment has provided an analysis of critical Information and Communication Technology (ICT) infrastructure in downtown San Diego. Fiber optic cabling, the backbone to ICT infrastructure in downtown San Diego, has been proven to establish critical links for essential private and government infrastructure to Internet and communication providers.

Additionally, the CARVER + Shock Information and Communication Technology infrastructure threat assessment has provided an objective approach to analyzing a critical infrastructure that is extremely subjective to both infrastructure contingency planners, and the common public. The CARVER threat assessment adaptation to ICT infrastructure will provide ICT threat assessment planners a tool to evaluate an often-forgotten critical infrastructure.

The CARVER Threat Matrix prioritizes targets by overall threat assessment score. This matrix provides an important view for City of San Diego homeland security planners. It organizes threat assessment targets into a numerical order, with the most vulnerable targets listed first. Government should address these most vulnerable targets first.

ANALYSIS OF THE THREAT ASSESSMENT

The analyses of the threat assessment results are intentionally vague with respect to specific downtown locations. This will ensure that this publicly available thesis does not offer a "roadmap" for a terrorism attack against San Diego.

The CARVER + Shock threat assessment of downtown San Diego's ICT infrastructure finds many commonalities within the underground fiber optic infrastructure. The agglomeration of fiber optic lines is the most prevalent around major downtown streets, and large office "skyscraper" buildings. Several intersections have every available fiber optic provider traveling through its right of way. These locations provide attackers a single avenue for causing the most damage to ICT infrastructure with a single attack.

Intersections with multiple providers and diverse fiber optic routes scored lower on the overall threat score. Several locations surrounding jails and major government centers only have a single provider. Similar to agglomeration, and with the existence of a single point of failure and no backup provider, this will prove to be disastrous.

Major ingress and egress routes are very apparent in the threat assessment results. Since a higher amount of bandwidth and communications traffic travels in and out through ingress and egress points, this would provide a logical choice for attack because of the difficultly to reroute and rebuild infrastructure that is a focal point of the network. Even more disturbing, the five providers utilize raised concrete bridges as ingress and egress for downtown San Diego. In the event of a natural disaster or a terrorist attack it may take significant effort and time to erect new non-essential bridges, especially if other life-saving priorities take precedent (see fig. 5).

45

Due to the sensitive nature of this figure,

it has been suppressed for general publication.

Figure 5. Five fiber optic providers use five different bridges as ingress and egress routes for Downtown San Diego.
Due to the sensitive nature of this figure,
the source has been suppressed for general publication.

Two publicly available right of ways exist from the downtown San Diego mainland to North Island, an island with a significant militarily presence. If critical military objectives exist on this island, the publication of ocean trench right of ways provide two agglomerated locations for attack to any cable reaching the island (see fig. 6).

Due to the sensitive nature of this figure,

It has been suppressed for general publication.

Figure 6. Two public right-of-ways to North Island.
Due to the sensitive nature of this figure,
the source has been suppressed for general publication.

FIBER OPTIC NETWORKS
(RECAP OF CHAPTER 2)

Chapter 2 details the background and importance of fiber optic networks and network providers. Network and fiber providers overbuilt their fiber optic networks during the dot-com boom in the late 1990's. Fiber optic cabling became the main solution for high-speed communication and data transport because of its extremely large capacity and low cost over long distances. Most copper POTS (Plain Old Telephone System) systems were replaced with communications systems that work over VoIP and fiber optic infrastructure.

The domination by fiber optics has lead to several vulnerabilities that exist in ICT infrastructure. Network agglomeration, the tendency for fiber optic providers to build fiber optic networks along routes utilized by existing networks, has created a single point of failure, exacerbated during natural disaster or terrorist attack.

Private sector efficiencies create vulnerabilities due, in part, to cost saving strategies by both private and public organizations. In order to reduce costs, fiber optic providers inadvertently create single points of failure.

Fiber optic networks do not just singly provide Internet connections. Fiber optics also provide data, voice, emergency 9-1-1 service, wireless cell phone tower connections, and even supplement microwave radio routes. The location information for fiber optic networks in the City of San Diego are available by using the state-wide DigAlert service, or by viewing construction permits in the Maps and Records section.

Conventional explosive weapons are the most logical choice for an attack on ICT infrastructure, specifically fiber optic networks. Asphalt, concrete and dirt are the only sources of protection for underground fiber optic cabling, from aboveground threats. The depth of asphalt and concrete are publicly available from the City of San Diego for each type of roadway.

CARVER + SHOCK THREAT ASSESSMENT (RECAP OF CHAPTER 3)

In Chapter 3, I explained the CARVER + Shock method for performing threat assessments. The Information and Communication Technology Infrastructure variation of the CARVER + Shock model uses a customized criterion for scoring potential threat aspects. This customized criterion is better adaptable to the specific time and infrastructure requirements of ICT infrastructure. For example, a twenty-four hour loss of activity for a business may not affect overall operations; however, an ICT infrastructure outage for only a few minutes may prove devastating.

I also conducted the threat assessment for ICT infrastructure in downtown San Diego. A CARVER + Shock threat matrix was constructed to better visualize the importance of a target's score on each of the evaluation criteria.

CITY OF SAN DIEGO POLICY RECOMMENDATIONS

The City of San Diego has no central, regularly updated repository of fiber optic installations. There is no policy which prohibits or encourages fiber optic carriers to build fiber optic lines a certain distance from each other, or on separate streets. Permitting processes for new fiber optic line construction in the City of San Diego are not congruent to establishing policy for maintaining a redundant network throughout all private data carriers.

This thesis underscores that the City of San Diego must create an enforceable policy, specifically for fiber optic installations, which defines the relationship and responsibilities between the City of San Diego and individual fiber optic providers. San Diego must immediately establish a centralized repository or database to store location information for fiber optic providers. This will ensure that fiber optic providers do not build near installations from other providers, possibly creating agglomeration, and a single point of failure.

A solution developed by Canadian graduate students Mark Tulloch, B. Eng., and Wensong Hu, BASc to address fatalities in Canada from inadequately marked utility line explosions, uses an asset management system combined with a GIS mapping solution to organize utility installations. The proposed tool would allow municipalities to collect information on existing utility line installations using mobile GIS units. The information would then be entered into a central database repository maintained by the city. New utility installations would be entered into the system and run through a corresponding algorithm for the specific type of installation. For example, one algorithm could check the distance between other fiber optic providers, or even be set to avoid agglomeration problems (see Appendix F).

The City of San Diego must not recommend solutions that create high levels of agglomeration. Rather, the city's policy should discourage agglomeration of fiber optic lines. This is not the current policy, as demonstrated in the City of San Diego authored memorandum listed in Appendix C.

Restrict access to information on utility line locations. California law requires construction companies to notify DigAlert (Underground Service Alert of Southern California) forty-eight hours before underground digging occurs. DigAlert then notifies companies that have utility line installations in the area. Utility line companies dispatch "locators" which mark their line location using flags, paint, and other markers.

Anyone posing as a construction or boring company can submit a utility line locator request form directly to DigAlert. This is an easy way to have utility companies identify the exact locations of utility lines, without having to conduct any research or recognizance to find the line locations on the part of a terrorist. Lines that may not be publicly available, such as those owned by the federal government, would also have to be marked.

The Underground Service Alert of Southern California must restrict the ability to submit utility line locator requests to authorized users only. This service could potentially be used against utility providers. DigAlert must verify each request is valid before marking lines.

Availability of "D" Sheets in the Maps and Records Division of the City of San Diego Development Services Department provide abundant utility line location information. "D" Sheets are publicly available permit request records. With enough research, "D" sheets can be used to accurately plot any utility line installation in the City of San Diego, including fiber optic lines. Even extremely vulnerable "switching stations" (locations that provide critical switching, power, and conditioning to communications lines) can be found using this method of recognizance. "D" Sheets should not be publicly available because they provide very detailed information on installation locations of dry utilities, including fiber optics.

Public and private building blueprints are available in the Maps and Records Division of the City of San Diego Development Services Department. Blueprints for communication facilities, fiber optic switching stations, government buildings utilizing ICT infrastructure, and public service installations are also publicly available in the City of San Diego, Maps and Records Division. Current restrictions do not provide any security for patrons that sign in under "Self-Help." Anyone could easily steal blueprints or microfilm, since the Maps and Records Division is an antiquated single room without proper surveillance. The only security precaution for blueprints requires copyright authorization for photocopies of any copyrighted blueprints. This is, of course, is if the blueprint is not photographed or taken first.

Fiber optic installations that require public approval should not be readily accessible on government websites. AT&T's NexGen fiber optic network from California to Texas traveled through Federal Bureau of Land Management public land. Due to the permitting process, AT&T was required to publish detailed specifications and route diagrams in order to pass an Environment Assessment. These detailed networks maps are still readily accessible from the Bureau of Land Management (see Appendix G).

FEDERAL POLICY RECOMMENDATIONS

"The United States has a policy on telecommunications; the policy is not to have a policy," says Jim Hayes, of the San Diego-based Fiber Optic

Association, Inc. Hayes insists that after the 1996 FCC deregulation CLECs (competitive local exchange networks) dominated the local telecommunications market. This created competition for the larger telephone companies such at AT&T, and promoted lower prices and increased technology for consumers. This same move, however, created an ever-increasing gap in centralized record keeping. [Record keeping] "is extremely expensive and difficult," says Hayes. "The phone companies have been very poor at keeping records; cities are sometimes no better, especially older cities."

Hayes suggests that the federal government should require municipalities to control record keeping, and charge the associated record keeping expenses back to the utility companies who use the public right of ways. As I stated in Chapter 2, utility companies recognized by the California Public Utilities Commission are not charged any fees by the City of San Diego.

The Federal government should recognize that private fiber optic carriers and government must communicate. Recognizing that both private fiber optic carriers and government institutions must communicate caused the Federal Communication Commission (FCC) to recharter the Network Reliability and Interoperability Council (NRIC) (FCC, par. 12).

The NRIC membership has been expanded to include wire line providers, as well as wireless, satellite, and Internet Service Providers (ISP). Homeland security and inoperability security concerns should be a primary focus of the NRIC.

A MILITARY VIEW OF
FIBER OPTIC NETWORK SECURITY

Excerpt from "Taking Down Telecommunications" by Major Gerald R. Hust, USAF:

Physical attack of the configuration is achieved by conventional, nuclear, or non-lethal weapons (destructive non-lethal attacks may include EMP, high voltage surge weapons, etc.). The main consideration in physical attack is 'the extent of damage which can be done to remote portions of the network from a localized attack.' System redundancy, centralization of key nodes, and hardness and location of those nodes will determine the resources required to obtain the desired system degradation from physical attack. Dispersal, hardening of key facilities, and rules of engagement all act to limit the effectiveness of conventional attack. For example:

> The fiber optic network Saddam Hussein used to communicate with his field commanders also included many switching stations (one of which was at the basement of the Ar-Rashid Hotel) and dozens of relay sites along the oil pipeline from Baghdad through Al-Basrah to the south of Iraq. However, hitting some of these targets was not desirable despite their military significance, because of possible collateral damage.

Deciding to use physical force does not relinquish the requirements for detailed intelligence. If a system is quite sophisticated, it is important to know the location of key nodes and their back-ups. Even if the attacker has the necessary intelligence, he must be willing and able to expend the effort to attack the system, at the expense of other potential targets. A halfhearted attempt may do little to degrade combat operations. If the attacker:

> has (and is willing to expend) enough ordinance, he can destroy all of the network communications assets (provided he can find and target them, and render the communications network inoperable. Short of this extreme, however, there are two key questions that the topological susceptibility assessment must address relative to the threat of physical attack:

1) Are there any nodes that control the entire network. One must also look beyond the immediate network to see if there is a key node providing vital information to another network such as a weapon system.

2) Is it possible to locate and prioritize these key nodes to insure maximum results from each bomb dropped. (24–25)

How vulnerable to a terrorist attack is downtown San Diego's ICT infrastructure? Downtown San Diego is a center for commerce, entertainment, residential living and government operations. Many of these critical organizations and operations provide services outside of San Diego to other parts of the state, nationally, and even internationally. Several vulnerabilities identified within this thesis need to be addressed immediately in order to ensure the survivability of critical operations in San Diego. The target threat assessment provided identifies vulnerabilities to downtown San Diego's ICT infrastructure.

How reliable are existing connections to ICT infrastructure? Existing connection reliability is displayed in the CARVER + Shock Threat Matrix. Connection reliability varies by location, number, and configuration of fiber optic providers.

Does the City of San Diego fiber optic installations policies ensure redundancy? As demonstrated in the federal and local policy recommendations section of this chapter, there are many recommended changes to both local and federal policies. To ensure redundancy, these recommendations should be adapted to existing policy.

FINAL THOUGHTS

This thesis has created fiber optic network threat assessment criteria previously unavailable in both private industry and in government. It established a baseline for evaluating threats to infrastructure that are difficult to visualize or map. Not only has it created a list of potential vulnerabilities and critical infrastructure problems, but it has also suggested solutions to solve these publicly known vulnerabilities.

In sum, agglomeration and cost-saving practices by private fiber optic providers, combined with inadequate or non-existent public policies, have created severe vulnerabilities in downtown San Diego's Information and Communication Technology infrastructure. This thesis has revealed an ever-increasing nationalized problem. Only with the cooperation and coordination between private businesses, fiber optic providers, local, state, and federal governments will we sufficiently protect our communication and information sharing systems, as well as our homeland.

WORKS CITED

Bell Canada Enterprises. "Public Safety Jeopardized as Aliant's Network Damaged." June 9, 2004. Nov. 25, 2005 <http://www.bce.ca/en/ news/releases/aliant/ 2004/06/09/71313.html>.

Berinato, Scott. "The Truth About Cyberterrorism." *CIO Magazine* Mar. 15, 2002. Nov. 25, 2005 <http://www.cio.com/archive/031502/ truth.html>.

Caloyannides, Michael A. *Potentially Catastrophic Vulnerabilities of the Internet and Proposed Remedies.* Falls Church, VA: Mitretek Systems, 2002.

Carson, Louis J. "Food Security Assessment: USDA Fruit and Vegetable Industry Advisory Committee." PowerPoint presentation. July 12, 2005. Dec. 3, 2005 <http://www.ams.usda.gov/fv/fviac/ 2005/carson.ppt>.

Center City Development Corporation. "San Diego Downtown Community Plan." 2005. Nov. 25, 2005 <http://www.ccdc.com/planupdate/ pdf/01_SDCP_Intro.pdf>.

City of San Diego. "Memorandum to Dry Utility Companies." Dec. 14, 2000. Nov. 26, 2005 <http://www.bandwidthbay.org/map/cityltr.htm>.

---. "San Diego Municipal Code, Article 2: Public Rights-of-Way and Land Development." Oct. 2001. Nov. 25, 2005 <http://clerkdoc.sannet.gov/RightSite/getcontent/ local.pdf?DMW_OBJECTID=09001451800acb8e>.

---. "San Diego Municipal Code, Article 2: Public Rights-of-Way and Land Development." Sept. 2003. Nov. 25, 2005 <http://clerkdoc.sannet.gov/RightSite/getcontent/ local.pdf?DMW_OBJECTID=09001451800aca0c>.

---. "Standard Drawings." July 1, 2004. Nov. 25, 2005 <http://www.sandiego.gov/engineering-cip/pdf/csdsd2003.pdf >.

Clarke, Richard D. "LNG Facilities in Urban Areas." May 2005. Dec. 3, 2005 <http://www.projo.com/extra/2005/lng/ clarkereport.pdf>.

The Conference Board of Canada. "Implications of the Terrorist Attack for the Canadian Economy." Sept. 19, 2001. Nov. 26, 2005 <http://www.conferenceboard.ca/press/ documents/ cboc-spreport.19.01.pdf?>.

Converge! Media Ventures, Inc. "Bell Canada Suffers Fiber Sabotage in Eastern Canada." June 10, 2004. Nov. 26, 2005 <http://www.convergedigest.com/Bandwidth/ newnetworksarticle.asp?ID=11388>.

Corley, W. Gene. "Applicability of Seismic Design in Mitigating Progressive Collapse." July 10, 2002. Nov. 25, 2005 <http://www.nibs.org/ MMC/ProgCollapse%20presentations/Corley's%20revised%20paper. pdf>.

Dixon, T. H. "The Rise of Complex Terrorism." *Global Policy Forum* Jan. 15, 2002. Oct. 30, 2005 <http://www.globalpolicy.org/ wtc/terrorism/ 2002/0115complex.htm>.

Downtown San Diego Partnership and Centre City Development Corporation. (n.d.). "Bandwidth Bay—Downtown San Diego." Nov. 26, 2005 <http://www.bandwidthbay.org/main.htm>.

Federal Communications Commission. "Remarks of Michael K. Powell Chairman, Federal Communications Commission at the NSTAC XXVII Executive Session Luncheon." May 19, 2004. Nov. 25, 2005 <http://hraunfoss.fcc.gov/edocs_public/attachmatch/ DOC-247404A1.doc>.

Federal Emergency Management Agency. "Risk Management Series: Risk Assessment—A How-To Guide to Mitigate Potential Terrorist Attacks Against Buildings." *FEMA* 452 Jan. 2005. Nov. 25, 2005 <http://www.fema.gov/pdf/fima/452/fema452.pdf>.

Gorman, Sean P. "Networks, Complexity, and Security: The Role of Public Policy in Critical Infrastructure Protection." Diss. George Mason University, 2004.

Green, Joshua. "The Myth of Cyberterrorism." Nov. 2002. Nov. 25, 2005 <http://www.washingtonmonthly.com/features/2001/0211. green.html>.

Grid Today. "How Does Optical Networking Effect the Grid?" Mar. 24, 2003. Nov. 26, 2005 <http://www.gridtoday.com/03/ 0324/101210.html>.

Hayes, Jim. Telephone interview. Nov. 7, 2005.

Hust, Gerald. R. "Taking Down Telecommunications." Master's thesis. School of Advanced Airpower Studies, Maxwell Air Force Base, Alabama, 1993. Nov. 25, 2005 <http://www.maxwell.af.mil/au/ aul/aupress/SAAS_Theses/SAASS_Out/Hust/ hust.pdf>.

Macy, Terry. "Fiber Optics Basics." Mar. 20, 1997. Nov. 25, 2005 <http://oak.cats.ohiou.edu/~sl302186/fiber.html>.

MathPages. "Infinite Parallel Redundancy." n.d. Dec. 3, 2005 <http://www.mathpages.com/ home/kmath326.htm>.

Mobiledia Corp. "Cell Phone Tower Search." n.d. Dec. 3, 2005 <http://www.cellreception.com/towers/towers.php?page=1>.

The National Academies—Computer Science and Telecommunications Board. "Growing Vulnerability of the Public Switched Networks: Implications for National Security Emergency Preparedness." 1989. Nov. 25, 2005 <http://www.nap.edu/books/ NI000389/html/>.

National Infrastructure Institute, Center for Infrastructure Expertise. "CARVER2™." (n.d.). Dec. 2, 2005 <http://www.ni2cie.org/ Carver2Flyer.PDF>.

Office of the Vice President—The White House. "Reengineering Through Information Technology-Part I." Sept 1, 1993. Oct. 16, 2005 <http://clinton6.nara.gov/1993/09/ 1993-09-01-npr-on-reengineering-through-information-technology-part.html>.

Queen, Cynthia. Personal interview. Sept. 30, 2005.

Sempra Energy. "Press Release—California Utility Agency Approves 'Last Mile' Telecom Solution." Oct. 3, 2003. Nov. 25, 2005 <http://public.sempra.com/newsreleases/viewPR.cfm?PR_ID= 1585&Co_Short_Nm=SE>.

Tulloch, Mark, and Wensong Hu. "A Proposed Solution for Mapping Underground Utilities for Buried Asset Management." 2005. Nov. 25, 2005 <http://www.ryerson.ca/ORS/showcase/ Mark_Tulloch_3rd_prize_2005.pdf>.

U.S. Census Bureau. "Information and Communication Technology: 2003." June 2005. Nov. 26, 2005 <http://www.census.gov/prod/ 2005pubs/ict-03.pdf>.

Wikipedia. "Dark Fiber." 2006. Nov. 26, 2005 <http://en.wikipedia.org/wiki/ Dark_fibre>.

Wireless Consumers Alliance, Inc. "Cell Towers—How They Work." 2004. Nov. 25, 2005 <http://www.wirelessconsumers.org/site/ pp.asp?c=giJYJ3OOF&b=14577>.

World Information Technology and Services Alliance. "WITSA Global Research Shows World's Largest Consumer of Technology Increased Spending Less Than 1% Last Year." Feb. 28, 2002. Nov. 25, 2005 <http://www.witsa.org/press/ DP02release.htm>.

APPENDIX A

"D" SHEET

Due to the sensitive nature of this appendix,
It has been suppressed for general publication.

Due to the sensitive nature of this appendix,
It has been suppressed for general publication.

Source: Due to the sensitive nature of this appendix, the source has been suppressed for general publication.

APPENDIX B

CITY OF SAN DIEGO, PROCEDURES FOR WORK ON UTILITY INSTALLATIONS IN THE PUBLIC RIGHT–OF–WAY

Article 2: Public Rights–of–Way and Land Development

Division 11: Procedures for Work on Utility Installations in the Public Right–of–Way
(Retitled from "Procedures for Use of Public Rights–of–Way by Public Utilities"
on 10–8–2001 by O–18995 N.S.)

§62.1101   **Purpose and Intent**

It is the purpose and intent of this Division to provide policies and procedures for the use of the public rights–of–way within the City in order to:

(a)   Preserve the public rights–of–way by conserving the limited space available within the public rights–of–way within the City.

(b)   Maintain safe conditions for the public use of the public rights–of–way within the City.

(c)   Minimize the inconvenience to the public.

(d)   Provide specific guidelines for the coordination of placement of installations to ensure a level of street improvement that is functionally safe, and preserves the integrity of public facilities.

(e)   To establish cost recovery mechanisms for inspections.
("Purpose and Intent" added 5–28–1996 by O–18309 N.S.)

§62.1102   **Definitions**

For purposes of this Division the definitions in Section 62.0102 apply. In addition, the following definitions apply:

"Blockage Report" means a report made on a City–created form indicating the area and type of work to be done in the public rights–of–way.

"California Coordinate System" means the coordinate system used to establish horizontal control, based on the North American Datum of 1983 (NAD83), as established by the National Geodetic Survey pursuant to Public Resources Code sections 8801–8819.

Ch.	Art.	Div.	
6	2	11	1

64

"Cathodic Protection" means control of external corrosion on underground or submerged metallic systems. "City's Standard Drawings" means that document on file in the Office of the City Clerk as Document No. 769819.

"Decorative Surface" means any non–standard surface on the public rights–of–way such as ceramic tile, concrete pavers, stamped concrete, or other surface using a unique treatment.

"Dry Utilities" means all public utilities other than those providing water, gas and sewage services.

"Inspection Fee" means the fee assessed pursuant to Section 62.1107 to reimburse the City for its costs of inspecting work in the public rights–of–way within the City.

"Installations" means any type of structure, apparatus, plant, equipment or other property installed in the public rights–of– way.

"Joint Trenches" means the mechanism approved by the City Engineer for the shared use by Dry Utilities of limited space in the public rights–of–way.

"Markout" means a marking on the pavement that identifies the type and approximate horizontal location of underground installations.

"Pavement" means the fully improved roadway surface within the public rights–of–way, designed and constructed to support the movement of vehicular traffic. Pavement typically consists of asphaltic concrete or Portland cement concrete.

"Pothole" means a limited excavation used to determine the actual (vertical and horizontal) location of underground installations.

"Trenching" means the type of excavation for the placement of installations in the public rights–of–way in accordance with City's Standard Drawings.

"Trench Plate" means a temporary structural steel plate, secured in place, to safely support legal loads over excavations in the public rights–of–way.

"Underground Service Alert" means the state–mandated agency responsible for, after receiving notice of a planned excavation, notifying all public utilities that have underground installations in the public rights–of–way prior to any excavation.

"Unimproved Rights–of–Way" means City rights–of–way that do not have pavement and do not have a sidewalk, curb or gutters.

("Definitions" added 5–28–1996 by O–18309 N.S.)

§62.1103   Authority of City Engineer and Duty to Obtain City Engineer Approval

The City Engineer is authorized to adopt procedures to implement this division. All persons shall obtain written authorization from the City Engineer before commencing any work on public rights–of–way within the City.
("Authority of City Engineer and Duty to Obtain City Engineer Approval" added 5–28–1996 by O–18309 N.S.)

§62.1104   Records

(a)   All persons with installations in the public rights–of–way shall maintain accurate records relating to the location of that person's installations. For this purpose the person shall use the California Coordinate System or the current system used by the person, providing that such system can be readily understood by others. Such records may not be relied upon to provide information other than the approximate location of the person's installations.

(b)   Within fifteen (15) days of receipt of a request, all persons shall make these records available to the City.
("Records" added 5–28–1996 by O–18309 N.S.)

§62.1105   Installations

(a)   All persons wishing to work in the public rights–of–way shall first call for markout, then pothole, whenever any excavation in the public rights–of–way makes it necessary to know the exact horizontal and/or vertical placement of that person's installations.

(b)   All such persons shall give Underground Service Alert a minimum of two (2) working days advance notice before any markout or pothole is commenced.

(c)   If unforeseeable circumstances arise requiring immediate action, marking–out and potholing shall be done within twenty– four (24) hours after Underground Service Alert is notified.

(d)   All cuts shall be made with a sawcut, rockwheel or other method approved by the City Engineer.

Ch.	Art.	Div.	
6	2	11	3

(e)   New installations placed in the public rights–of–way shall occupy the locations indicated in the City Standard Drawings unless otherwise authorized by the City Engineer.

(f)   All installations placed in the public rights–of–way shall comply with City's Standard Drawings.
(Amended 10–8–2001 by O–18995 N.S.)

§62.1106   Placement and Removal of Markouts

(a)   Markouts shall not be placed in the public right-of-way more than thirty (30) days prior to the commencement of excavation work performed in connection with an installation. If the excavation work is not commenced within thirty days of the placement of the markout, the markout shall be immediately removed.

(b)   Markouts shall be removed from all surfaces in the public right-of-way, including decorative surfaces, within thirty (30) days of the completion of the excavation work, if the work is completed, but in any event no later than sixty (60) days from the date the markout is placed in the public right-of-way.
("Placement and Removal of Markouts" added 10-8-2001 by O-18995 N.S.)

§62.1107   Documents Required for City Engineer Authorization

(a)   All persons placing installations in the Public rights–of– way shall file a Blockage Report with the City Engineer no later than two (2) working days prior to commencing any work. After review of the Blockage Report, the City Engineer may require that person to file a traffic control plan.

(b)   For any installations funded by a public utility, other than lateral installations or other minor installations as determined by the City Engineer, at least two (2) months prior to beginning any cut, the public utility shall submit to the City Engineer copies of maps which indicate the area and location of installations. For any installations funded by public utility customers, the public utility shall provide the requisite number of copies of maps as soon as such installation is planned.
("Documents Required for City Engineer Authorization" renumbered from Sec. 62.1106 on 10–8–2001 by O–18995 N.S.)

§62.1108   Inspection Fees; Inspections

(a)   All persons placing installations in the public rights–of– way shall pay an inspection fee to the City Engineer. The City Engineer has the authority to set the schedule of fees collected pursuant to Section 62.1108 provided that such fees do not exceed the reasonable cost of conducting the random inspections authorized by Section 62.1108(f). The inspection fee will be used to reimburse the City for the costs of conducting the random inspections set forth in Section 62.1108(f).

(b)   The inspection fee shall be paid either: (1) prior to each inspection, or (2) by making payment to the City within thirty (30) calendar days of having received an invoice from the City. Invoices will be sent by the City no more frequently than on a monthly basis.

(c)   If a person elects to make an annual deposit, upon request by the City Engineer, the person shall deposit additional money when the funds on deposit are exhausted.

(d)   Funds on deposit shall be carried forward from year to year until expended.

(e)   If a person makes a payment pursuant to Section 62.1108(b) (2), the City Engineer will, within thirty (30) days of receipt of payment, provide that person with a copy of the field reports from inspections and a detailed accounting of the number of City Staff hours performed on the inspections.

(f)   The City Engineer may conduct random inspections of any work being done in the public rights–of–way, based on information provided in the Blockage Report. The City Engineer may inspect the work for compliance with all applicable laws, ordinances and construction standards with emphasis on the following:

(1)   Traffic control procedures.

(2)   Compliance with City street restoration standards.

(3)   Compliance with the pavement cutting procedure.

(g)   If a City inspection discloses nonconformance with any of the requirements of this Division, the City shall provide written notice of the nonconformance within five (5) working days. The person placing the installation shall implement the corrective work specified by the City Engineer within five (5) working days of receipt of written notice of nonconformance. If the corrective work is not completed within five (5) working days of receipt of written notice of nonconformance, the City may perform the necessary repairs and all costs related to the repair shall be charged to the person placing the installation.

(h)   The City Engineer shall have the authority to stop work and to request that the excavation be uncovered to certify compliance with this Division.

(i)   Any City work done directly or indirectly to ensure compliance with the provisions of this Division shall be charged to the person placing the installation which requires the City work.

(j)   Any work done which is the result of a City required project shall be exempt from the inspection fee requirement of Section 62.1108.
("Inspection Fees; Inspections" renumbered from Sec. 62.1107 on 10–8–2001 by O–18995 N.S.)

§62.1109   Pavement Restoration

(a)   All persons excavating in the public rights–of–way shall restore pavement at the end of each of each day with either temporary or permanent pavement.

(b)   If permanent surfacing material cannot be installed within forty–eight (48) hours, by the end of each day all intersections, pedestrian crossings and other locations as required by the City Engineer shall be trench–plated or backfilled such that the excavation may be driven upon by vehicular traffic.

(c)   All damaged pavement shall be restored with surfacing material which matches both the surface and the structural strength of the adjacent surface.

(d)   All pavement on the public rights–of–way shall be restored with permanent surfacing material within seven (7) days where there are more than two lanes of travel, and within thirty (30) days where there are two or fewer lanes of travel.

(e)   Any striping removed or temporarily placed shall be restored within twenty–four (24) hours where there are more than two lanes of travel, and within seventy–two (72) hours where there are two or fewer lanes of travel.
("Pavement Restoration" added 5–28–1996 by O–18309 N.S.)

(10-2001)

§62.1110   **Restoration of Decorative Surfaces**

In addition to the requirements of Section 62.1108, in any area where there is a decorative surface on the pavement:

(a)   Before disturbing any decorative surface, all persons excavating in the public rights–of–way shall provide information to the City Engineer to establish that it is necessary to disturb the decorative surface because other alternatives, such as rerouting, boring, jacking or scoping, cannot be used.

(b)   Before commencing work on the decorative surfaces, specifications shall be prepared that are designed to minimize destruction and ensure restoration of the same quality of surface. The specifications shall be submitted to the City Engineer for approval.

(c)   Written notice shall be delivered to the City Engineer at least two (2) working days before starting construction or trenching that will involve any disturbance of decorative surfaces. The notice shall include the location and estimated start and completion dates.

(d)   If unforeseeable circumstances arise requiring immediate action, written notice shall be delivered to the City Engineer as soon as possible upon the start of construction.

(e)   If the unforeseeable circumstances requiring immediate work arise after normal business hours, written notice shall be delivered to the City Engineer at the beginning of the next regular working day.

(f)   If it is necessary to remove any decorative surface, it shall be removed without damaging adjacent surface material.

(g)   In the public rights–of–way in the Centre City area of City, removable sections shall be designed and installed over any installations involving a decorative surface to provide access to the installations without destroying the decorative surface.

(h)   Decorative surfaces shall be restored, at no cost to the City, with surfacing material that matches both the surface and the structural strength of the adjacent surface.

("Restoration of Decorative Surfaces" added 5–28–1996 by O–18309 N.S.)

Ch.	Art.	Div.	
6	2	11	7

§62.1111   **Safety and Traffic Control**

(a)   All persons working in the public rights–of–way shall be responsible for the safe movement of both vehicular and pedestrian traffic through that person's construction and maintenance operations.

(b)   The City Engineer shall be notified of scheduled construction at least two (2) working days before commencing work.

(c)   Signs, warning devices, traffic control plans and general conditions of safety, as described either in the City's Standard Drawings or other State standards, shall be maintained.

(d)   All persons performing work in the public rights–of–way shall identify him, her or itself with on–site signs indicating the name of the person and the phone number to call in case of a complaint or emergency. Such signs shall remain on–site for two (2) weeks after completion of work.

("Safety and Traffic Control" added 5–28–1996 by O–18309 N.S.)

§62.1112   **Relocation of Installations**

(a)   All persons maintaining installations in the public rights-of-way shall relocate or remove their installations whenever such relocation is necessary for a proper governmental purpose, whether or not that purpose is to be accomplished by a public entity or by a private entity on behalf of a public entity. In such cases, the cost of the relocation or removal shall be borne by the person.

(b)   When installations need to be relocated or removed as a result of construction by a private entity, except as set forth in Section 62.1112 (a), the cost of such relocation or removal shall be borne by the private entity undertaking the construction. That private entity shall contact the owner of the installations affected by the work to advise them of proposed improvements. That private entity shall also make specific arrangements for the relocation of any conflicting installations.

(Amended 4-12-1999 by O-18632 N.S.)

Ch.	Art.	Div.	
6	2	11	8

71

§62.1113   Cathodic Protection

(a)   Public utilities maintaining installations in the public rights–of–way shall
       provide Cathodic Protection in accordance with the practice of the National
       Association of Corrosion Engineers (NACE).

(b)   If the NACE standards conflict with either the California Department of
       Transportation or California Public Utilities Commission's requirements, the
       most stringent requirements shall govern.

("Cathodic Protection" added 5–28–1996 by O–18309 N.S.)

§62.1114   Quality Control

(a)   All persons performing work in the public rights–of–way are solely
       responsible for ensuring that the work performed, whether by that person,
       contractors, subcontractors, employees, agents or representatives, complies
       with all applicable City and State standards.

(b)   At the beginning of each calendar year, each public utility with installations in
       the public rights–of–way shall submit a quality control plan and emergency
       closure plan to the City Engineer for approval.

       (1)   The quality control plan shall indicate the number of inspectors and
              the areas to which they are assigned.

       (2)   Each emergency closure plan shall indicate steps to be taken during a
              flood or earthquake to address safety issues.

       (3)   An emergency closure plan shall be filed with the City Engineer and
              with the City Emergency Operations Center.

("Quality Control" added 5–28–1996 by O–18309 N.S.)

Ch.	Art.	Div.	
6	2	11	9

APPENDIX C

MEMORANDUM TO DRY
UTILITY COMPANIES FROM
THE CITY OF SAN DIEGO

Ballpark District, Fiber Optic Networking Opportunity

THE CITY OF SAN DIEGO

December 14, 2000

To: Interested Dry Utility Companies

The City of San Diego is in the process of pursuing implementation of an ordinance which imposes a moratorium preventing excavation in newly constructed or resurfaced streets for a number of years from the date of construction or resurfacing. Currently, a large redevelopment effort, involving a number of different private parties as well as public entities, is ongoing in the East Village area of downtown San Diego which will result in a large number of newly constructed or resurfaced streets. As a result, it may be difficult or more costly for excavations to occur in that area upon completion of the redevelopment effort.

The City would like to take this opportunity to encourage all companies who may desire to install facilities in the streets in the East Village area to begin planning where and when they may need facilities in order to avoid delay, or to avoid being impacted by the anticipated moratorium.

In addition, the City has become aware of an ongoing effort by private entities, to coordinate planning and construction of a joint trench to share in the costs and expenses of design and construction of new facilities in the streets of East Village. Because there is a limited amount of available space in the streets in the East Village area, the City is encouraging every effort to plan ahead and to coordinate future construction. Although this joint trench is not a City project and any involvement is voluntary, participation in this joint trench may provide efficiency and realization of cost savings for those involved. Any company interested in obtaining more information about participation in this joint trench may send a written notice to:

Maroun El-Hage, P.E.

Associate Civil Engineer
Engineering & Capital Projects Department
1010 Second Avenue, Suite 1200
San Diego, CA 92101
Fax: 619-533-3071
e-mail: mce@sdcity.sannet.gov

The notice should contain the following information:

• Name address and contact information

• CPCN number

• Nature of utility service

• Interest in a joint trench vs. a separate facility

• Anticipated number, size, and type of required conduits.

For reference maps of the proposed redevelopment, please visit the following web sites:

www.bandwidthbay.org and www.ccdc.com

Sincerely,

PATTI BOEKAMP
Acting Director
Engineering & Capital Projects Department

Transportation & Drainage Design Division

Engineering & Capital Projects • Public Works • 1010 Second Avenue, Suite 1200 • San Diego,
CA 921014905

Tel (619) 533-3173

Source: City of San Diego. "Memorandum to Dry Utility Companies."
Dec. 14, 2000. Nov. 26, 2005 <http://www.bandwidthbay.org/map/
cityltr.htm>.

APPENDIX D

MAPS OF CELL PHONE TOWER INSTALLATIONS IN DOWNTOWN SAN DIEGO, CALIFORNIA

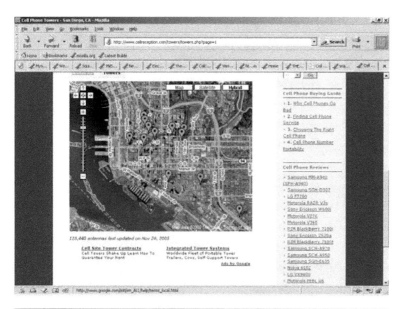

Overview of cell phone towers deployed in the downtown San Diego area. An example of the vast information you can obtain from public sources, about cellular phone towers in downtown San Diego.

Source: Mobiledia Corp. "Cell Phone Tower Search." n.d. Dec. 3, 2005
<http://www.cellreception.com/towers/towers.php?page=1>.

APPENDIX E

CARVER + SHOCK
THREAT ASSESSMENT

Due to the sensitive nature of this appendix,
it has been suppressed for general publication.

Source: Due to the sensitive nature of this appendix, the source has been
suppressed for general publication.

APPENDIX F

PROPOSED SOLUTIONS FOR CITY OF SAN DIEGO RECORD KEEPING SHORTFALLS

A PROPOSED SOLUTION FOR MAPPING UNDERGROUND UTILITIES FOR BURIED ASSET MANAGEMENT

Mark Tulloch, B. Eng., MASc. Candidate
Wensong Hu, BASc., MASc. Candidate

E-mail: {*mtulloch@ryerson.ca, whu@ryerson.ca }

Introduction

The management of buried infrastructure is at the forefront of activities of many municipalities. Since visual inspection of the underground utilities is not possible, it is hard to determine or estimate when to rehabilitate or replace utilities. Underground asset management applications assist municipalities in managing their buried infrastructure. Accurate asset management is heavily reliant on accurate information. Determining the location of buried utilities is one source of information that constitutes good asset management practices. This article presents a solution for accurately mapping the exposed underground utilities, and for creating an underground utility database that holds all of the necessary information pertinent to underground asset management.

The focus of our research is the development of a mapping system to enable the acquisition, assembly, manipulation and management of spatial data defining the location of underground utility services. This information will improve utility and transportation infrastructure management activities and assist municipalities with managing the subsurface space in the road allowance, particularly in an urban environment.

Problem Identification and Objectives

There are two problems concerning underground utilities that are gaining more and more attention in Canada. The first is that the construction risk of underground utilities is increasing due to the uncertainty of the location of the utilities. The ideology of 'out of sight, out of mind' cannot apply to underground utilities any longer. This problem is leading to project delays, extra work orders, change orders, construction claims, contingency bidding, loss of service, property damage, and worst of all, injury and death. On April 23, 2003, seven people were killed in an explosion in Etobicoke, Ontario due to a roadwork crew accidentally puncturing a gas main. The Ministry of Labour and the Technical Standards and Safety Association have laid charges against a number of contractors and sub-contractors for the erroneous delineation of the gas main on the utility map. Just five days after the Etobicoke explosion, a worker was killed, and three more injured, in a gas main diversion job in Windsor, Ontario. Accurate mapping has the potential to improve many activities associated with the management of underground utilities and other assets in the right of way, including improved safety and cost avoidance during utility and road reconstruction.

---

* Corresponding Author

1

The second problem concerning underground utilities is the inadvertent expenditure of public money on infrastructure upgrades due to the lack of, or unreliable, information concerning underground utilities. The municipality of Halton reported that it has spent an average of $10 million per year on replacing underground sewer and water pipes in the past eight years. Upon inspection of the old pipe, all sections were found in excellent condition, and could have lasted another 20 years. The replacement of these water and sewer pipes was decided using erroneous data from engineering drawings prepared in the late 1800's.

The overall objective of this project is to provide location information of underground infrastructure to allow for increased public safety, and for better underground asset management. The solution to this problem requires field data collection of utility information using both photogrammetry and the global positioning system (GPS). A geospatial information system (GIS) is also required to store and organize the utility information, so that accurate asset management can be achieved my municipal engineers. The remainder of the article will focus on how these tools will be used to accurately collect, store, and analyze underground utility information.

Proposed Solution

The proposed solution for accurately mapping underground utilities for utility management requires many components. Figure 1 describes the relationships of these different components. Utility management can be divided into two main categories: mapping and asset management. The thrust of the research at Ryerson will be concerned with utility mapping, but there will be some overlap with the asset management side of the flow diagram (see Figure 1). The central tool that will be used to organize the mapping information is a mobile GIS. Two technologies will be used to map the exposed utilities. The first and most predominant method for mapping underground utilities is photogrammetry. The photogrammetric method provides three deliverables to be stored in a GIS: the utility coordinates, imagery, and attribute information. The coordinates of the utility features are the most critical information, and they will be stored in conjunction with   the   attribute   information   of   the

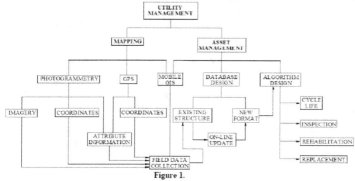

Figure 1.

utility features. The captured imagery will also be stored, and may be employed at a later date to provide further information.

The second method to map underground utilities is by GPS. The proposed GPS data collection method would provide two sources of information that would be stored in the GIS. The utility features coordinates could be derived from the GPS survey, as well as attribute information about the utility features. Both the photogrammetric method, and the GPS method, would be controlled through a mobile GIS, which would allow an easy information exchange between the field data collection and the GIS.

Asset management is the other half of the solution that is being addressed in this paper. The last component to the proposed solution is the algorithm that will be used to determine the four questions of asset management:

- What is the cycle length of the utility?
- When should inspection occur?
- When should rehabilitation occur?
- When should replacement occur?

It is important to realize that the algorithm design for asset management purposes is out of the scope of this project, but it is also important to have a comprehensive concept of how the algorithm derives answers to these four questions, and what information is needed to answer the four questions. By determining what information is necessary to asset management, the database design can incorporate all of the required information to perform this task.

Expected Results

To evaluate the effectiveness of this project a set of three criterions is used. The first is the efficient transition of digital information from the data collection to the utility database. There are many different types of data being used to map underground infrastructure including GPS signal data, photogrammetric images, existing raster images, control files, and user input information. There is a great necessity to ensure that all of the gathered information can be easily transformed from the data collection to the utility database. Not only does the information need to be complete, but also it must be organized in such a method that easy visualization can occur.

The second criteria that will be used to measure the effectiveness of this project is the spatial mapping accuracy of the photogrammetric and GPS mapping methods. To evaluate this criteria conventional ground surveying will be used to map the underground infrastructure to an accuracy of 1-4 cm. Both the photogrammetric and GPS mapping methods will be compared to the ground survey to determine the accuracies of the respective methods. The photogrammetric method is expected to provide accuracies of 10-20 cm, and the GPS method is expected to achieve accuracies of 5-10 cm. Both of these methods yield results within the limits of asset management systems.

The third and final criterion used to measure the effectiveness of this project is the user-friendliness of the mapping system. One of the goals of this project is to develop a mapping system that allows for a user to have little knowledge about photogrammetry or GPS. This eliminates the need for highly trained personnel to operate the mapping system, which is consistent with the desires of municipalities. To evaluate this criterion, close-working relations will develop between the Ryerson research team and different

3

municipalities in order to facilitate constructive dialog to improve the user-friendliness of the mapping system. By meeting these three criterion, the mapping system will be effective, which will permit accurate mapping of underground infrastructure, and therefore underground asset management practices will improve.

Acknowledgements

This project would not be successful without the support of the City of Toronto. The purchase of the related mapping equipment is greatly appreciated. We would also like to acknowledge our supervisor, Dr. Mike Chapman, for his continued assistance and participation in this project.

4

Source: Tulloch, Mark, and Wensong Hu. "A Proposed Solution for Mapping Underground Utilities for Buried Asset Management." 2005. Nov. 25, 2005 <http://www.ryerson.ca/ORS/showcase/Mark_Tulloch_3rd_prize_2005.pdf>.

APPENDIX G

AT&T NEXGEN PUBLICALLY
AVAILABLE NETWORK MAPS

Due to the sensitive nature of this appendix,

it has been suppressed for general publication.

Source: Due to the sensitive nature of this appendix, the source has been suppressed for general publication.

ABSTRACT OF THE THESIS

The Vulnerability of Fiber Optic Networks:
A CARVER + Shock Threat Assessment for the
Information and Communication Technology
System Infrastructure of Downtown San Diego,
California
by
Lance William Larson
Master of Science in Interdisciplinary Studies: Homeland Security
San Diego State University, 2006

It is imperative that fiber optic cabling, arguably the most important of our nation's critical infrastructures, is protected from terrorist attacks and other devastating threats. By utilizing a combined protection effort between national, state, municipal, and private industry, and setting appropriate and enforceable policy for fiber optic lines security and installation, a better strategy for infrastructure protection can be achieved.

Fiber optic cabling, the backbone of information and communication technology (ICT) infrastructure, carries critical data transmissions, phone conversations, and financial transactions from business and government in downtown San Diego to the outside world. Without fully operational fiber optic cabling, business and government operations would grind to a halt.

Agglomeration (the tendency for fiber optic providers to build fiber optic lines closer and closer to one another), and private sector efficiencies, such as building a single trench to encase fiber optic cabling in order to save money, are main contributors to fiber optic vulnerabilities. The City of San Diego also contributes to fiber optic vulnerabilities by relying on fiber optic installation providers to self-regulate, and by promoting the use of shared trenches to minimize construction and pavement disturbances. Finally, location information for fiber optic lines in downtown San Diego city streets is widely available via the Internet, and directly from the City of San Diego using publicly available sources.

Using the CARVER + Shock threat assessment model, originally developed by the military as a targeting assessment tool, ICT infrastructure targets in Downtown San Diego were analyzed and prioritized in a 1500-page threat assessment. Targets with a higher threat score should immediately be evaluated by terrorism planners.

The results of the threat assessment showed several policy shortfalls and lack of policy for fiber optic line installations, several locations where fiber optic cabling agglomerates in critical government and business centers, and several proposed single points of failure.

The results of the study will be disseminated to homeland security agencies and to local governments.